SPATIAL EFFICIENCY IN GEOGRAPHY

The Authoress

Surekha Pandit Bapat voluntarily retired as a lecturer in geography from the Government V.N. Institute of Arts and Social Sciences (formerly, Morris College), Nagpur. She has been a meritorious student throughout and has won prizes and medals during her career. She was also a holder of National Scholarship. She has presented short research papers in national and international geographical conferences and Indian Science Congress, and has to her credit a number of published papers in geographical journals as well as in edited volumes on geographical topics. Also, she has a book to her credit on Geography for primary students published on behalf of Balbharati, and books on Geographical topics for University students.

Dr. Surekha Pandit Bapat earned her Ph.D. on Spatial Efficiency in Geography in 1991 under the guidance of Prof. N.G. Jain.

SPATIAL EFFICIENCY IN GEOGRAPHY

DR. (MS.) SUREKHA PANDIT BAPAT

CONCEPT PUBLISHING COMPANY, NEW DELHI-110059

Cataloging in Publication Data—DK
Courtesy: D.K. Agencies (P) Ltd. <docinfo@dkagencies.com>

Bapat, Surekha Pandit.
Spatial efficiency in geography / Surekha Pandit Bapat.
p. cm.
Includes bibliographical references (p.)
Includes index.
ISBN 8180692825

1. Geography. 2. Land use—India. I. Title.

DDC 915.4 22

ISBN 81-8069-282-5 (H.B.)

First Published 2006

Published and Printed by
Ashok Kumar Mittal
Concept Publishing Company
A/15-16, Commercial Block, Mohan Garden
New Delhi-110059 (INDIA)

Phones : 25351460, 25351794
Fax : 091-11-25357103
Email : publishing@conceptpub.com

Dedicated
to
My Parents

PREFACE

Geography has always been a spatial science, looking for a spatial order of things in this chaotic world just as other disciplines look for an order in their own fields. However, it remained for a long time largely descriptive at the expense of objective analysis and generalisations. The word "geography" itself means the description of earth and that is a common belief about geography. Therefore, a common person like Alice 'through Looking Glass' makes a grand survey of the country and says 'it is something very like learning geography... principal rivers, principal mountains, principal towns...'. With the revival of the locational approach, the basic spatial nature of geography came into focus, leading to various spatial themes and theories and giving it a solid scientific footing. The quantitative revolution in geography helped in the process quite a lot by objective methods of measurement and analysis. The development of geographical discipline is an ongoing process and the concept of spatial efficiency is a small contribution to it.

Spatial efficiency is a general tendency of man in his spatial behaviour, arrived at by trial and error. It works as a controlling principal of his activities in space. As such, it could be measured and also simulated for planning. The work in this direction has started recently. This book is devoted to elaborate discussion of the principal of spatial efficiency and its practical application to various problems of human geography.

31-08-2003

Surekha Pandit Bapat

ACKNOWLEDGEMENTS

I would like to thank from my heart all those who helped me directly or indirectly in completion of this work. First of all, my thanks are due to my guide and inspiration Prof. N.G. Jain and Prof. V.H. Deodhar for their valuable suggestions and encouragement. Dr. Mrs. S.S. Deshpande and Prof. V.B. Changole helped me in preparation of the text and I owe them sincere thanks. Mrs. D.S. and Mr. S.V. Deodhar deserve to be thanked for producing computer copy of the work. I also express my gratitude for the cooperation of the publishers without which this book would not have come out.

31-08-2003 **Surekha Pandit Bapat**

CONTENTS

1

INTRODUCTION

Geography studies physical and human phenomena and their spatial organisation which is not haphazard. On the other hand, it is controlled by certain spatial principles. On the basis of such principles, spatial themes and models are built up in geography to provide rational interpretation of areal differentiation. Spatial efficiency is one of them which is able to offer geographical explanation of many human activities in space.

Man strives for spatial efficiency in his activities or spatial behaviour, knowingly or unknowingly. To be efficient in relation to space is advantageous and desirable for him. To understand the concept of spatial efficiency, a word about the meaning of space in geography will not be beside the point.

The Earth's Space: A Core Concept of Geography

Geography is a spatial or chorological science just as history is chronological. Space in geography is a deeper concept than merely answering the 'where of things'. It is a basic organising concept of geographic information and methodology and a unifying force for all its branches.

In his passion to comprehend his own world, man said to himself

'The Earth is rude, silent, incomprehensible at first, be not discouraged, keep on—'

(Walt Whitman, cited by Woolridge and East, 1970, p. 35). And he has kept on, creating a host of sciences and disciplines to set the reality in different perspectives. In this gamut of sciences, geography has a spatial perspective. It is primarily concerned with the spatial analysis of phenomena. Says Chisholm (1968, p. 4), 'whereas the hallmark of historical studies is the integration of data with respect to time, in geography the distinguishing characteristic is concern with (terrestrial) space' Haggett (1968, p. 13) has also called a concern for location and distribution in space as a hallmark of all geographical writing and a theme in methodological reviews.

The concept of space is the very core of geography in which it finds its own ground where it is most fully itself. All the geographical branches are imbedded in the concept of space and become justifiably geographical. Whatever the topic of study, geography essentially has a spatial concern.

The spatial nature of geography was recognized very early by Immanuel Kant, even though the location school of geographical thought now upholds it anew. Kant was very clear in stating that history studies the phenomena which follow one after the other in time while geography studies them as they lie in space. The earth's space is not void—it is full of phenomena, physical and cultural, making up the environment and forming the subject matter of geography. Or rather, space is conceived by the presence of various environmental phenomena, and system of relations among these phenomena. Notwithstanding, these relations of phenomena have a spatial aspect which is really the central theme of geographical study and research (Lowental, 1975, p. 105). Since the real interest of geography is the spatial aspect of phenomena which can be well comprehended by their visual representation, maps have become its indispensable tool. Map is a common technique for all branches of geography and it also promotes internal unity among them all.

The living reality of space is marked by its general coherence achieved through spatial organization of phenomena. Diverse phenomena are taking place and making the earth's space look complex and chaotic. 'The earth's space in which we live is one of infinite variety and no two areas are precisely the same' (Thoman and Corbin, 1974, p. 158). Even the same area changes through time. There are constant flows of commodities, people and ideas. Everything is in flux. But there is an order in chaos which the geographer looks for. The variable and complex phenomena are spatially interrelated and organized. Relief is a function of structure and climate. A town is interlinked to its hinterland. An irrigation work depends on terrain and drainage and affects population and economy. In sigwart's words, there is more order in the world than appears at first sight, but that is not discovered till the order is looked for (Haggett, 1968, p. 2). A geographer's task is to investigate this spatial organization of phenomena or coherence of space, which gives rise to his fundamental belief that there is unity among all kinds of diversity on the surface of the earth.

The view of space as a basic organizing concept in geography first inclined to the Newtonian notion of absolute space, having an existence of its own and containing environment and its phenomena. Much of the philosophy of geography stems from a 'container' view of space, associated with the concepts of Newton and Kant. According to Kant, space is not a thing or event but a framework for things and events to get arranged. This view is largely discredited since the discovery of non-Euclidean geometries as topology of advanced mathematics. The idea of relative space is rooted in modern relativity theory of Einstein in which one's awareness of space is based on a perceptual extension of the location of phenomena. The relative space means the existence of space relative to objects or phenomena and the space is defined by the relationships between them. This concept of relative space may be regarded as a useful advancement upon the abstract idea of absolute space.

In relativistic terms, phenomena and space are inseparable from each other, but the sequential arrangement of phenomena itself forms a spatial frame for man to refer to. Both space and time may be described as a frame of reference which is intuitively used in experience.

The spatial nature of geography is explicit in some of its well-known definitions calling it the study of: (i) places (Preston James) (ii) space (Kant) (iii) Areal patterns of co-variation (Robinson, Lindberg, and Brinkman (iv) Location, distribution, interdependence and interaction in place (Luckermann) (v) The man-earth system (Berry) (vi) rational interpretation of areal differentiation (Hartshorne). The most widely accepted definition of geography by Hartshorne (1959, p. 21) states, 'Geography is concerned to provide accurate, orderly, and rational description and interpretation of the variable character of the earth surface'. He seems to have followed Hettner (1905) who also referred to the variable character of the earth surface and the interrelations of these variations in phenomena. Thus, the concept of space signifies the geographical approach.

Meaning of Geographical Space

In geography, space is not looked upon as outer astronomical space, the infinite and ubiquitous void of the universe except for fixing the celestial position of the earth. Rather, geographical space is finite and restricted to the earth's surface with a few kilometers just below and above it. In general, the moho discontinuity may be taken as its lower limit and the atmospheric boundary as the upper one (Cole and King, 1968, p. 2). This is the major field of geographical studies of the complex sets of interrelationships that characterise the content of areas.

This space is three dimensional and full of phenomena—physical and cultural—which form the environment. So much so that the space itself is known by the presence of environment or one's awareness of space is subject to the perception of the environment. The environmental phenomena happen to be the living reality of space and invariably have spatial dimensions.

The expression of spatial dimensions which is predominant in any geographical discussion, is in terms of location, distance, direction and area (Kolars and Nystuen, 1974, p. 1). Their dimensional qualities are point location-no d, distance and direction-1d, area: 2d and areal volume: 3d. They are represented on maps likewise by points or dots (no d), lines or links (1d), patches or surfaces (2d) and spheres or blocks (3d). Such drastic abstractions are commonly made in geography and are an essential part of cartography. (Cole and King, 1968, p. 9). Various combinations of these dimensions are also possible like dots on line, patch on patch etc. In addition, there is the fourth dimension of time and temporal analysis is also important in geography.

Spatial or Locational Analysis

All geographical analysis is done within the spatial frame of reference. A geographer is really interested in spatial implications of phenomena in the form of spatial patterns and distributions, spatial behaviour and interrelationships which have spatial dimensions. Berry (1964, pp. 2-5), opines that the integrating concepts of the geographer relate to spatial arrangements and distributions, to spatial interaction and organization and to spatial processes. Mc Carty (1954, pp. 96-97) also proclaims that a geographer analyses spatial patterns of man-environment relationships. Geographical literature revolves round spatial or locational analysis of phenomena.

Spatial analysis leads to spatial themes and models to offer geographical explanation. Though geography is largely descriptive, and does not much venture into the explanation of genesis and mechanism of phenomena, it seeks spatial principles behind the evolution of spatial patterns and interrelationships in the analysis of them. Spatial efficiency is such a principle which is able to offer explanation of many human activities in space. If human spatial behaviour is analysed, it will be found to follow the tendency towards spatial efficiency. In this book, spatial efficiency has been developed as a spatial theme and model useful as an analytical tool in various geographical problems.

The Scheme of the Book

The purpose of this work is to bring out the underlying principle of spatial efficiency in human activities and to develop an operational model to measure, predict and plan it wherever possible. The book is divided into 10 chapters discussing spatial efficiency theoretically and applying its model in various geographical problems.

Chapter 1 gives a bare outline of space as a core concept of geography and introduces the principle of spatial efficiency. It also takes stock of all the chapters that would follow.

Chapter 2 is fully devoted to develop the concept of spatial efficiency with its definition, implications and model building.

Chapter 3 deals with the Central Place Theory of Christaller to show how spatial efficiency holds good there. The theory has been applied to Vidarbha region as an example and validity of spatial efficiency is discussed. An example of simulation with the help of spatial efficiency model is also presented.

Chapter 4 deals with the structure of route networks—their

connectivity, accessibility and detouring and discusses the influence of spatial efficiency. The analysis of route network is incomplete without consideration of movement. In spatial flows, efficiency is sought in movement minimization as well as the maximization of traffic. A simulation model of spatial efficiency in movement is applied.

Chapter 5 is about location of industries. Efficiency is utterly required in the process of manufacturing. Nonetheless, it is necessary in respect of the location of industrial plant. This has been a topic of discussion for economists and economic geographers for a long time. It has been considered from spatial efficiency point of view.

Chapter 6 discusses the location of boundaries and shape of regions. The boundaries of administrative or political units are laid as physical as well as cultural divides. They form the shape of the regions which should desirably be efficient. They also involve the location of the headquarters of the regions they bound.

Chapter 7 deals with the urban land use patterns, which are geographically interesting. Again, a clear distinction has to be made between the Western and the oriental, especially the Indian urban land use. The famous urban land use models are discussed in the light of spatial efficiency.

Chapter 8 has the topic of agricultural land use for discussion. Agriculture creates specific patterns of land use which are mainly influenced by natural set of conditions on the one hand and the economic conditions on the other. The farmer strives to be spatially efficient as much as possible among them. The classical Thunenian model is discussed in this respect.

Chapter 9 deals with perception in geography. Perception analysis has been a recent development in geography, but it throws a new light on man-environment relationship. Spatial efficiency has to be perceived by man for his decision-making and preferences. However, there is much scope for further research in this regard.

Chapter 10, the last one is a short conclusion of the whole work.

REFERENCES

Berry B.J.L (1964), Approaches to Regional Analysis: A synthesis, *Annals of the Association of American Geographers*, 54, pp. 2-5.

Chisholm M. (1968), *Geography and Economics*, G. Bell And Sons Ltd., London, p. 4.

Cole J.P and C.A.M. King (1968), *Quantitative Geography*, John Wiley and Sons Ltd., London, p. 2.

Haggett. P. (1968), *Locational Analysis in Human Geography*, Edward Arnold Ltd., London, p. 13.

Ibid, p. 2.

Hartshorne R. (1959), *Perspectives on the Nature of Geography*, John Murray, London, p. 21.

Kolars J.F. and J.D. Nystuen (1974), Geography, *The Study of Location, Culture and Environment*, McGraw Hill Book Company, New York, p. 1.

Lowenthal D. (1975), Geography, Experience and Information: Towards a Geographical Epistemology, in E. Jones (ed.), *Readings in Social Geography*, Oxford University Press, London, p. 105.

Mc Carty H.H (1954), *An Approach to a Theory of Economic Geography*, Economic Geography, Worcester, pp. 96-97.

Thoman R.S and P.B. Corbin (1974), *The Geography of Economic Activity*, McGraw-Hill Book Company, New York. p. 158.

Woolridge S.W. and W.G. East (1970), *The Spirit and Purpose of Geography*, Hutchinson University Library, London, p. 35.

2

THE CONCEPT OF SPATIAL EFFICIENCY

Spatial efficiency has been recently developed, not yet fully, as a spatial theme to tackle socio-economic problems in geography. A few authors used it in their empirical problems as a technique but it was not developed as a full fledged concept. It is really a meaningful concept and a useful technique of geographical data handling. In this sense, it goes to form a part of ideological structure of the discipline of geography. Such spatial theories and analytical techniques lead us to new insights about the real world and suggest how the facts should be arranged. The combination of geographic facts, techniques and theories establishes a geographic point of view (Kolars and Nystuen, 1974, p. 8)

Spatial efficiency is an underlying principle in human activities of various kinds. Environment offers man a certain number of choices and leaves him free to decide between them. He is guided in his choice by the principle of spatial efficiency and the ensuing behavioural pattern is an outcome of his endeavour to attain it. It is the task of the geographer to find how and how much spatial efficiency is achieved in his endeavour, with a good operational model for measurement.

The model of spatial efficiency implies optimisation and becomes a normative model. It is not only useful to measure efficiency but also to simulate it. It is, therefore, quite worthwhile in planning and proves to be a handy tool in dealing with various spatial problems.

Meaning of Efficiency

The word efficiency signifies maximum work with minimum effort. So, it is a maximum-minimum or maximin concept. Efficiency is attained with maximisation of work but minimization of effort or time at one and the same time. Brisco (1914, p. 33) says that from the most primitive times, men under the natural inherited stimulus have always sought to obtain a desired result with the least possible efforts. Efforts are saved so that all sorts of waste are eliminated. However though, maximisation of work is striven for, efficiency does not at all mean strenuousness, which demands extra efforts. On the other hand, efficiency brings greater results with less efforts. In modern world, efficiency has became a watchword to achieve progress and development. Planning for development, therefore, must consider efficiency as a vital factor.

The formula for efficiency is expressed as:

$$\text{Efficiency} = \frac{\text{Maximisation of work}}{\text{Minimisation of Efforts}}$$

The equation shows that efficiency increases in proportion to the work done but in inverse proportion to the effort. Thus, there is an inverse relationship between work and effort and it operates at one and the same time. Higher the value more will be the efficiency.

Definition of Spatial Efficiency

In geography, efficiency is taken into account with reference to space. Man is generally inclined to make most of the available spatial potential while putting in as less effort as possible in traversing distances. Thus, spatial efficiency may be defined as a combination of maximisation of function or activity and minimisation of distance or transportation at one and the same time. Smith (1968, p. 137) states that efficiency usually implies minimisation of distance or transportation costs. The efficiency criterion or objective function can however, involve maximisation of some quantity such as the flow over a given route or through a specified terminal.

Spatial efficiency, then, involves two parameters: (i) maximisation of some function or activity as nominator and (ii) minimisation of distance as a denominator. Spatial efficiency is the ratio between the two. It increases in proportion to activity but in inverse proportion to distance and the highest ratio is the optimum. Either of these two factors is inadequate to determine spatial efficiency on its own, unless the other is reasonably held constant. If one computing spatial efficiency considered the factor of distance minimisation alone without proper regard of the other factor, then the picture is lopsided. Truly speaking, spatial efficiency lies in the ratio of these two factors and the highest possible ratio happens to be the optimum efficiency. This optimum may lie at a level less than actual maximum and/or more than actual minimum values. The concern is not for actual maximum or minimum values but for the highest ratio between them. The maximisation and minimisation affect each other and the highest possible compromise between the two leads to spatial efficiency. For example, the straight least distance route may be spatially less efficient than a longer route which generates more traffic but it must not be too long so as to affect spatial efficiency adversely. Spatial efficiency is a delicate balance between the maximum and minimum and it is struck in between the two.

In his spatial behaviour, man follows the principle of spatial efficiency when he attempts to wrest as much advantages from the environment as possible and to put as less efforts in traversing distances as possible. It is a common knowledge that man always inclines to cut short the distances and to maximize his gains in spatial activity. In principle, spatial efficiency is like gravity or potential models which also use the proportion between population and distance. Wilson (1970) also suggests that the whole family of gravity and potential models can be derived from maximizing principle in empirical analyses of flows of goods, people, ideas or anything between places. He is actually suggesting the principle of spatial efficiency. The gravity and potential models use only population as maximizing factor. But the concept of spatial efficiency includes all kinds of spatial activities in it.

Man strives for spatial efficiency but to reach it, he has to perceive the appropriate way in which he can maximize his activity and at the same time, minimize the distance. Man as a decision-maker has a vital role to play in his endeavour to achieve spatial efficiency. Sometimes, his decision is not appropriate as his perception of maximizing and minimizing methods may be wrong and at other times, his endeavour is half-hearted or falls short. Efficiency is obtained first by acquiring knowledge and then making proper use of that knowledge. (Brisco, 1914, p. 10) Consequently, he may not reach the optimum in spite of his desire. Therefore, there arise several levels of spatial efficiency attained by man. Many times it so happens that man stops at satisfying level and the effciency he has attained is one of satisficer kind and truly not the optimum. Whatever the level of spatial efficiency, it is attained through a long process of trial and error over many years. The process goes on to achieve more and more efficiency, or optimacy.

The behavioural school of thought does not believe in rigid economic optimacy and for them the satisficer kind of efficiency is that man aims for. Therefore, they take recourse to the probabilistic models rather than deterministic. It is true that many times, the real optimum is hard to find because of some constraints on the one hand and the changing conditions on the other. Maximization or minimization may be only an approximate value. McCarty and Lindberg (1966, p. 12) opine that the generalizations and principles of economic geography always appear as approximations, and that the entire research effort in the field is aimed at increasing the accuracy of those approximations. More refined methods, modes and machines continually raise the standards of maximization of activity and minimization of distance. So, the optimum is bound to change. In that case, the optimum has to be restricted to the given set of conditions. However, different levels of spatial efficiency are very well brought out in camparitive studies. The comparative studies are also desirable to set the goal for achievement of spatial efficiency if the optimum is not available.

The principle of spatial efficiency is based on the spatial themes that: (i) there is a tendency to prefer shorter distances and to seek proximity, (ii) there is a strong distance decay element, (iii) there are central and agglomerative tendencies, (iv) there is competition for space in which comparative advantages are sought, and (v) improved technology increases efficiency. According to Harmansen (1971, p. 15) many students in the field of spatial economics agree that the basic factors which bear upon the basic structure of spatial organization can be summarised under these headings: the demand for space, indivisibilities, economics of scale and costs of transportation and movement.

Spatial efficiency can be measured with the help of its two basic parameters of maximization of spatial activity and minimization of distance. The measurement is helpful not only to throw light on the level of spatial efficiency attained but also to set the guideline for improvement. It cuts the long process of trial and error short and accelerates the process of development. Therefore, in area planning, measurement of spatial efficiency, and also its simulation can never be ignored.

Maximisation of Spatial Activity

Efficiency is necessarily related to maximisation of work and spatial efficiency also has the aim to maximise some spatial activity such as utilisation of land or resource, movement over space, distribution of people etc. It is measurable in some quantity like population, or volume of flows etc. Against this maximising factor, the minimising factor of distance has to be compared. The maximisation should not go too far to affect the minimising factor adversely. In some themes, this factor is presumed to be constant for brevity's sake or for idealisation and only the minimisation of distance is sought. But it cannot be ignored in consideration of spatial efficiency. In the same way, maximization of spatial activity cannot determine, by itself, spatial efficiency. It has to be compared against the minimization of distance which is another parameter of it.

Minimisation of Distance

Minimization of distance is coupled as a denominator with maximization of spatial activity in determining spatial efficiency. Distance minimization is, so to speak, a very popular geographical theme used in several spatial problems. McCarty and Lindberg (1966, p. 26) also remark that the minimum distance concept has a great thematic value in spatial generalizations.

Distance is minimized because there is friction of space in traversing distances which requires effort. There is a tendency of man to follow the law of minimum effort (Losch) or the principle of least effort (Zipf) in his activities, so that wasteful energy and time are saved as far as possible. Lloyd and Dicken (1968, p. 145) opine that the principle of least effort is, without doubt, a powerful influence on man's behaviour.

Thus, distance is a barrier and efficiency lies in minimizing its barrier effect. Therefore, the activities are so located as to lessen the friction of distance. However, distance is often spoken of in terms of cost or time and in geography, cost distance or time distance are commonly used rather than mere physical distance.

Model of Spatial Efficiency

All theoretical discussion of spatial efficiency should lead to building of an operational model which is practically useful in real world situations and which represents the concept of spatial efficiency. A simple model can be constructed with the help of its two parameters as the two variables, expressed in some quantity. Just like the equation of efficiency, it is a maximum-minimum model implying that spatial efficiency increases in proportion to maximising spatial activity or $SE \propto A$ but in inverse proportion to minimising distance or $SE \propto \frac{1}{d}$ simultaneously. The formula may be written as:

$SE = G\frac{A}{d}$ where,

SE = Spatial Efficiency; A = Spatial Activity and d = Distance. G is a constant which is assumed to be unity and hence ignored.

Both the maximum-minimum variables of spatial efficiency are interdependent. One cannot do without the other. Really speaking, spatial efficiency is a compromise between these two variables. In a hypothetical case (Fig. 2.1), for example, A has the central and therefore the least distance location. But spatially most efficient point is B to locate an activity that requires population as its clientele.

Assuming hypothetical distances and using the formula $SEi = \sum_{\substack{i=1 \\ j=1}}^{n} \frac{Pi}{dij}$

where SEi = Spatial efficiency of i th place,

P = Population; and dij = distance from i to j, the following spatial efficiency has been worked out for 3 places, A, B, and C. Higher the value, more is the efficiency.

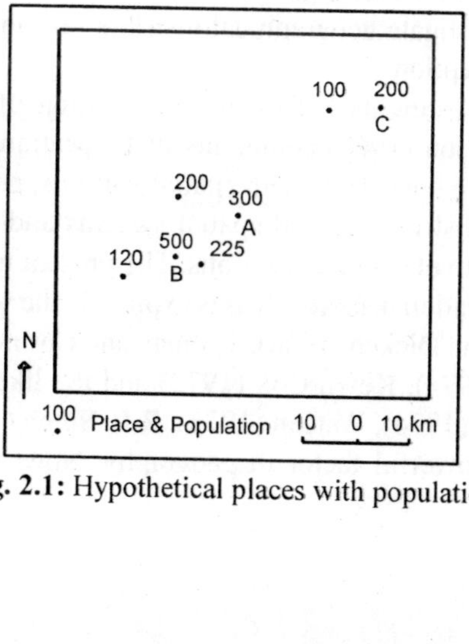

Fig. 2.1: Hypothetical places with population

Table 2.1: Hypothetical Spatial Efficiency

Place	*Population*	*Spatial Efficiency* $\left(\sum_{\substack{i=1 \\ j=1}}^{n} \frac{Pi}{dij}\right)$
A	300	86
B	500	210
C	200	35

The basic model may be moulded to suit the empirical problem, but its basic tenet does not change. The operational model is able not only to measure spatial efficiency but also to simulate it.

Development of Locational Theory and Efficiency

Geography and geographical thought began to develop appreciably in the interwar period, and many geographical concepts, principles and models had their origin in this period. It, then, became an applied science to be used for the planning purpose. After the Second World War, there had been a considerable shift in the conceptual framework of geography in the USA. (Hussain, 1990, p. 167). There was a shift towards social aspect from the physical side, to systematic specialization and ecosystem from regional analysis. Barrows propounded the concept of Human ecology; Jefferson and Saur favoured chorological themes. The period of 60s witnessed quantitative revolution in geography, making way for statistical methods to investigate geographical distributions and arrangements and spatial perception.

These developments led to the formulation of the locational theories. Dickinson (1969) comments that Locational Theory is an area in which the concern is with spatial patterns, geometry of flow linkages, spatial structures and spatial systems and with normative models leading to efficiency solutions. The present trend is certainly to the spatial-locational approach as is explicit in the works of Haggett (1968), Lloyd and Dicken (1968), Thoman and Corbin (1974), Kolars and Nystuen (1974), Kevin Cox (1972) and the like. Ramchandran (1980), Ullman (1974), Watson (1974), R.L. Singh (1989) discussed distance as a powerful factor in geography. Now, the geographers are using the advanced technology or remote sensing and aerial photos for interpretation of spatial phenomena.

In the Soviet Union, geographers helped the state planning in the selection of industrial location and resource development. The Soviet geographers, together with the experts of other sciences became important in regional planning, territorial industrial complexes, agricultural schemes, recreational and tourist facilities and specially in the selection of optimal sites for their construction and for routing of roads. There is a trend towards the purposeful transformation of the environment so as to optimise (Gerasimov, 1981, p. 32) man's living and working conditions.

The geographers of the 20th century had a feeling that they did not have theories and laws like other sciences and to overcome it, they started postulating their own structured ideas to interpret organizations in space. The earlier locational theories were borrowed from spatial economics. Christaller was the first geographer to make a major contribution to location theory. It really paved the way for the development of the ideas on efficiency.

Geography has passed out of descriptive and telelogical phase. Though most of the geographical laws are empirical in nature, the location theories give nomothetic (general) laws. Hussain (1990, p. 232) expresses the hope that new nomothetic laws would be formulated to put geography on a sound footing. The concept of spatial efficiency provides geography such a general principle of human spatial behaviour. It is not only explanatory but normative also. Therefore, like other locational theories, simulations and predictions are possible with the help of spatial efficiency model.

Positivistic point of view emphasises empiricism, while pragmatists hold spatial laws valid. Existentialism insists on a concern for human values, personal decisions, and human existence in space. Spatial efficiency is a valid spatial principle, uses empirical data and is an expression of general human tendency. It has a potential to give maximum-minimum solutions to human spatial problems and to make geography a result-oriented discipline.

Conclusion

The concept of spatial efficiency is the maximum-minimum concept, very useful in geographical enquiry. However, it has not been much employed as a practical method of analysis in spatial problems. Literature on spatial efficiency is scanty and much empirical work as well as its theoretical discussion is yet awaited.

The spatial efficiency measure has been tried for health centres in Salcette taluka of Goa by Hodgson and Valadares (1983), and by Massam and others (1984). Maurice Yeates (1974) used it in transportation problem of bussing school-children in Grant County, Wisconsin (USA). Morrill (1967) also presented a transportation problem for service areas of physicians. In most of the works, the emphasis is laid on minimum distance rather than its combination with maximum activity. Surekha Pandit (1984, pp. 526-530) discussed it with both the maximum-minimum variables.

The present trend in geography is hopefully towards the spatial or locational approach which emphasizes the spatial perspective of geography and shows a great concern for distance as a controlling factor of human activities in space. This concept would further develop into spatial efficiency if it is combined with maximization of spatial activity. It should, however be borne in mind that the optimum efficiency is the best possible in the existing conditions, liable to change if the present level of knowledge and technology is surpassed.

Walther Christaller (1933) used the minimum distance concept in his famous Central Place Theory. Actually he was discussing spatially efficient location of the central places and the shape of their tributary areas. The next chapter reviews the Central Place Theory and point out spatial efficiency in it.

REFERENCES

Brisco N.A. (1914), *Economics of Efficiency,* The McMillan Company, New York, p. 33.

Ibid, p. 10.

Gerasimov I. (1981), The Historical Background to Russian And Soviet Geography in *Soviet Geography Today*, Progress Publishers, Moscow, p. 32.

Hussain (M) (1990), *Evolution of Geographical Thought,* Rawat Publications, Jaipur, p. 167.

Ibid, p. 232.

Harmansen T. (1971), Spatial Organization and Economic Development: The Scope and Task of Spatial Planning, in R.P. Mishra (ed.) *Development*

Studies, No.1, University of Mysore, p. 15.

Kolars J.F. and J.D. Nystuen (1974), *Geography: The Study of Location, Culture and Environment,* McGraw-Hill Book Company, New York, p. 8.

Lloyd P. and P. Dicken (1972), *Location in Space: A Theoretical Approach to Economic Geography,* Harper and Row Publishers, New York, p. 145.

McCarty H.H and J.B Lindberg (1966), *A Preface to Economic Geography,* Prentice-Hall, Inc., Englewood Cliffs, N.J., p. 26.

Morrill R.L (1967), The Movement of Persons and Transportation Problem, in W.L. Carrison and D.F. Marble (eds), *Quantitative Geography, Part I: Economics and Cultural Topics, Studies in Geography*, 13, NW University Press, Illinois.

Pandit, S. (1984), Spatial Efficiency: A Point of View, *The Deccan Geographer*, Vol. 22, No. 3, Secunderabad, pp. 526-530.

Smith R.H.T. (1974), Concepts and Methods in Commodity Flow Analysis, in M.E.E. Hurst (ed.), *Transportation Geography*, McGraw-Hill Book Company, New York, p. 137.

3

SPATIAL EFFICIENCY IN LOCATION OF CENTRAL PLACES

Central places act as the centres of goods and services for the surrounding region and perform the task of spatial organization of service activities. These population clusters are a concrete expression of human occupance of space and their distribution creates spatial patterns. As such, they become an essential element of landscape and an interesting topic of study in geography.

The question that often perplexed the human geographers is whether the location and distribution of central places is systematic. At the outset of his treatise, Walther Christaller raises the question, 'Are there laws which determine the number, distribution and size of towns?' (Carter, 1972, p. 69). He worked on the problem with the simple belief that the population clusters would never occur haphazardly. Their location, size and spacing would show some relationship with each other and to the space around. Putting forward the deductive Central Place Theory, he arrived at regular hexagonal pattern of their distribution and sought its explanation in spatial economy *i.e.* minimisation of distance. On closer inspection, it will be found that the locational pattern of central places is really controlled by the principle of spatial efficiency *i.e.* a compromise between the maximization of the goods and services and minimization of distance.

The Central Place Theory

Walther Christaller produced in 1933 his well known theory of central places, giving a new insight into their location and distribution pattern. He presumed certain premises or ideal conditions such as isotropic surface and equal transportation facilities. Then he propounded that the central places were located centrally within the clustered communities at points of minimum distance for the populace. They carried out functions which the population of the surrounding area demanded. He envisaged their uniform distribution which should ideally produce hexagonal pattern with a higher order central place surrounded by six lower order central places. He recognised a whole hierarchy of the central places from the lowest to the highest on the basis of increasing threshold and range of goods and services, that determined the varying intensity of centrality. The size of the hexagonal service areas would differ accordingly and become nested into one another. A uniform lattice pattern with the central places equidistant from each other on the same level of hierarchy would result. (Fig. 3.1), August Losch (1954), a spatial economist, also arrived at the same conclusions while working on economics of location of the function of breweries. Christaller applied his theory to Southern Germany.

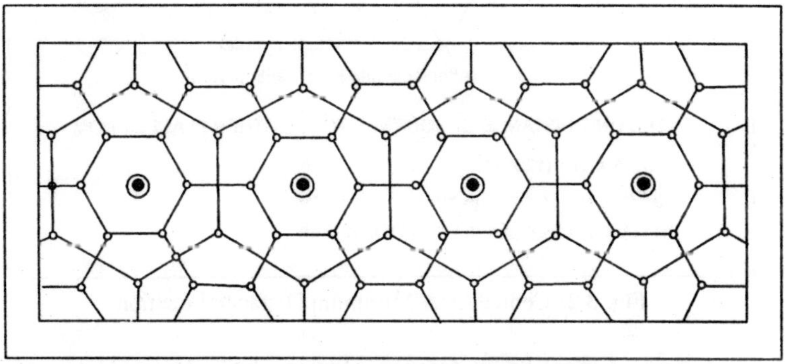

Fig. 3.1: Christaller's Central Place System

Minimum Distance Concept in Central Location

Both Christaller and Losch were setting forth the concept of spatial economy or minimum distance in central location of activity. A central location in an actually or conceptually bounded area happens to be a minimum distance position from all the sides and it requires minimum aggregate effort in travel, in keeping with the basic principle of least effort in human organization of activities.According to Cole and King (1968, p. 492), this is not only a point from where the average distance to all other places in the area is minimum, but one from where the maximum distance to the remotest place in the area is also minimum (Fig. 3.2).

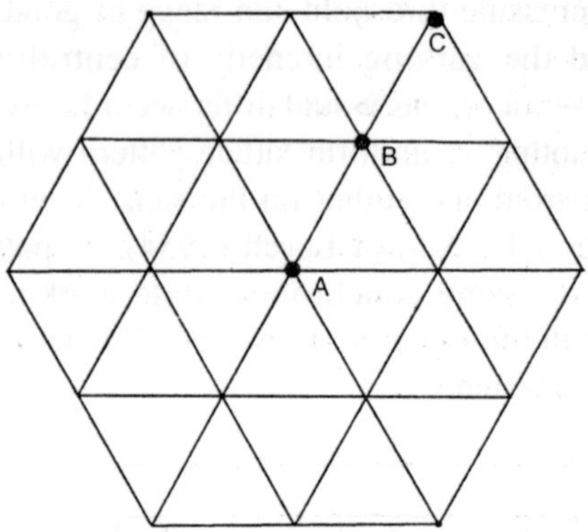

(After Cole and King, 1968, p.493)

THE UNIT DISTANCE OF JOURNEY TO ALL OTHER PLACES FROM :

A : (6 X 1) + (12 X 2) = 30

B : (6 X 1) + (7 X 2) + (5 X 3) = 35

C : (3 X 1) + (5 X 2) + (5 X 3) + (5 X 4) = 48

Fig. 3.2: Centre As A Minimum Distance Location

A central location in service area is, thus, desirable to offer goods and services in order to save distance for the customers. Taking a simple example in duopoly of two vendors on a beach (Alonso, 1968, p. 371) where population is assumed to be equally spread, it can be seen that the vendors, first located at random, will optimally move to the quarter points of the beach, or to the centre of their respective sale area, half each, so as to remain at minimum distance from their own customers. This Loschian solution of minimum distance in spatial competition is accepted as the efficient one. The classic economic model of H. Hotelling (1929) showed a clustering of both the vendors in the centre of the beach. However, in that case, they were actually on the p eriphery of their respective sale area which they carved out half each and the average distance for the customers was unnecessarily long. It could be reduced to the minimum if the vendors moved to the quartile positions. The Hotelling type of centre clustering cannot imply a stable locational equilibrium. The preference is for a Loschian solution. (Berry, 1967, p. 86).

In the similar way, a central place rural or urban, gets centrally situated in its hexagonal service area for the convenience of minimum travel for its clientele. The central tendency or the centrality becomes the essence of the central places following the minimum distance principle. B.J. Garner (1967, p. 304) also states that the locational decisions (of central places) are taken, in general, so as to minimise the frictional effects of distance.

Modern technology is effectively making movement easier and faster. Thereby, the service area of the central places is enlarging, and the modern times are witnessing tremendous growth of urban places, and their tributary areas. But the basic centrality or the control of the minimum distance principle remains unchanged.

Distortion of Regular Distribution

Christaller envisaged a very regular distribution of central places which were equidistant on the same level of hierarchy surrounded by six lower order central places. He concentrated on minimum distance and held equally powerful maximising factor constant. Had minimum distance been the only determinant of the locational structure of central places, the Christaller type regular distribution of them would evolve. But, in reality, it happens to be much distorted and Christaller's theory appears to be too ideal. Sometimes, this distortion is so high that validity of the theory itself has been doubted by workers like Dacey (1962) and Clark and Evans (1954). Empirical investigations revealed irregularities rather than regularities in areal distribution of central places. As Carter (1972, p. 115) puts it, if obeying the demands of threshold and range, particular grades of towns were marked by characteristic assemblages or bundles of functions, they seemed most certainly not to display any regularity in distribution. It tends to be more random than uniform. As a result, both size and shape of the hexagons lack uniformity.

The distortion from perfect regularity in the distribution of central places is due to the irregular distribution of population, their income and means of transport. The central places draw their clientele from the population which exerts unequal pull on the location of central places, and the regular lattice is correspondingly distorted. Taking the former example in duopoly of two vendors on a beach, and adding irregularity in the population distribution there, it can be seen that the vendors will be drawn to the more populous part of their sale area and they will be displaced from their quartile positions.

Spatial Efficiency in the Distribution of Central Places

Christaller's ingenious development of the concept of minimum distance in the location and distribution of central places earned an important place for his theory inspite of the deviations from the ideal. As Sumita Ghosh (1998, p. 65) comments, his concept of centrality, functional relationship, population thresholds and hierarchy are even today fundamental in the study of any settlement system.

Throughout his discussion of an ordered adjustment of settlements to the factor of distance to minimise movement, Christaller was actually putting up an argument for an efficient arrangement. In envisaging a hexagonal shape of the tributary area of a central place also, he adopted a spatially efficient shape. First, he accorded to it a circular shape which is spatially most efficient, but since the circular shape is impossible for the network in reality as the circles do not fit with each other leaving either a gap or an overlap, he converted the circular to the hexagonal shape of areas. Losch also converted the circles into hexagons of his economic regions. It is a proven fact that the hexagon is an efficient shape next to the circle in a lattice. Haggett, (1968, p. 49) says that the hexagon retains most of the advantages of the circle which is about four-fifths as efficient as the circle in terms of maximum radial distance and perimeter.

However, Christaller did not take into account the influence of the other maximising factor of spatial efficiency on the actual distribution of central places or rather held it constant in idealising his model. Therefore, empirical observations rarely came close to the pattern that he envisaged. But he was establishing a deductive theory which should be general and ideal. As Sumita Ghosh (1968,

p. 64) remarks, the spatial aspect of this theory is based on a delicate balance of several factors and even a slight change in one will alter the pattern like the location of resources, industrialization, mode and speed of transport and historical and political issues.

The maximising factor of spatial efficiency that distorts the regularity in the Central Place model is the distribution of population for whom all the goods and services in the central places are meant. The central places maximise their activity for the clientele they draw from the population of the surrounding area. Naturally, therefore, they will locate in the midst of more populous parts of the region. Since population distribution in any real world region is more often than not unequal, reflecting most of the inequalities in the region, the distribution of the central places is bound to be affected in response to the pull of demand from the population. As a result, their distribution appears distorted. However, it does not mean that it is spatially inefficient. On the other hand, it is really fulfilling maximization of activity of offering goods and services while remaining in the centre of population *i.e.* at minimum distance for the clientele. In applying Christaller's model to real world situation, the distribution of population should also be considered and distortions due to it should be taken as revealing the principle of spatial efficiency.

Doxiadis (1969, pp. 29-37) has developed settlement formation theory under five principles which call for optimization. The first two of them are just like the two variables of spatial efficiency *i.e.* (i) maximization of man's contacts with natural elements, people and cultural elements, and (ii) minimization of the effort. The other three are allied to these two and thus his ideas are not far from spatial efficiency. R.B. Mandal (1981, p. 373) also stresses that people crowd together in settlements in order to maximize the interaction at the spatial convenience centres and to achieve efficiency of central places at accessible points. Thus, both the maximum-minimum factors of spatial efficiency are equally important to explain the actual pattern of distribution of the central places in any region.

Due to irregular distribution of central places, which form hexagons around the higher order central place, not only the shape of the hexagons but also their size is distorted. It will be found that in a densely populated region, the hexagons are more regular and smaller as the central places are closer together, than in a sparsely populated one, in which the higher order central place in the centre will have to draw its clientele over a larger area. So also, it is not far from the fact, that the higher order central place is surrounded by less than six lower order central places and thus, it is not exactly a hexagon, because it is not a closely packed system of central places as Christaller idealized. There is much distortion in population distribution with varying income levels and varying means of transportation. Really speaking, the distortions in the shape and size of the hexagonal pattern of the central places speak of the factor of maximization of goods and services for the populace that the central places could attain in the given conditions. Spatial efficiency does not depend on the regularity in their distribution but on their centrality which, in turn is measured by the goods and services they offer to the people, remaining at minimum distance from them. Christaller's theory should, thus, be reviewed with spatial efficiency point of view.

A Case Study of Vidarbha

A town distribution of Vidarbha is produced as a case in point. (Fig. 3.4) Vidarbha, a part of Maharashtra State in India, is a predominantly agricultural region, a table land and a fairly populated area except in hilly tract and forests. For brevity's sake, a whole range of central places in rural country side is ignored and only the towns and cities as recognized by the census of India 1991 are considered for their hierarchy and distributional pattern. The census has its own classification and gradation of towns purely on the basis of population. It does not touch upon the centrality values of the towns which are very necessary to array them in the central place hierarchy.

According to 1991 census, there were 95 towns in Vidarbha catering to their clientele with goods and services, which should get due recognition in computation of the centrality values of the town. Therefore, urban workers in various categories of secondary and tertiary activities, urban literacy, commercial and industrial electric connections, educational facilities at different levels of schools, degree colleges, professional institutes and university and the facility of the banks and credit societies, representing demographic and socioeconomic conditions of the towns have been chosen as variables. A centrality index is prepared on the basis of location quotients (Toyne and New by, 1972, p. 105) derived for urban workers, urban literacy and electric connections. The educational and banking facilities are there to serve not only the local but also the population from the surrounding areas. The surplus population at each level of these facilities over and above the threshold population is worked out, and converted to percentage for each town. The composite indices are plotted against population of each town on a semilog paper which shows a gradation of towns in 5 classes. (Fig. 3.3) The hierarchy thus derived can be compared with the census classification and the difference noted. The following table gives both the classifications. The bases of the census classes and the hierarchical orders are quite different.

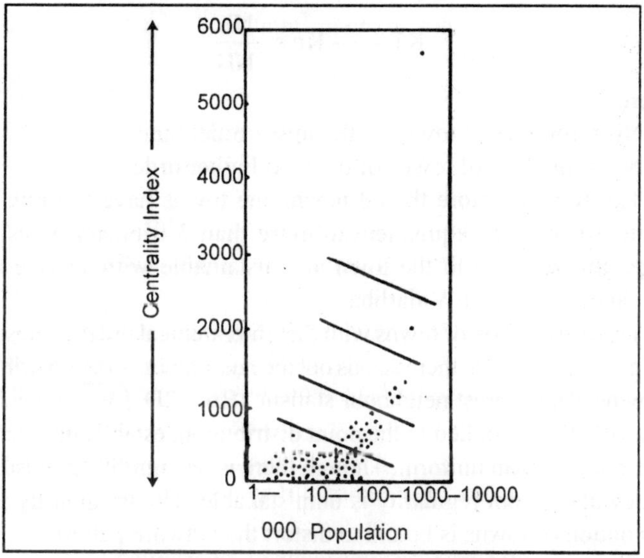

Fig. 3.3: Hierarchy of Towns in Vidarbha

Table 3.1: Hierarchy of Towns in Vidarbha

Town	*Census Classification*		*Hierarchical Orders*	
Class	*Population*	*No. of Towns*	*Centrality Index*	*No. of Towns*
I	1,00,000 & above	7	Above 2000	1
II	50,000-99,999	10	1501-2000	1
III	20,000-49,999	34	1001-1500	5
IV	10,000-19,999	30	501-1000	22
V	5,000-9,999	9	0-500	66
VI	Less than 5000	5		
	Total	**95**		**95**

In the functional hierarchy, the lowest order consists of the largest number of towns, while at the top, there is only one big city, the million city of Nagpur. The number of towns in each order is slightly different from that given by Christaller from the Ist to the Vth order as 1, 2, 6, 18, 54 for the market principle K = 3. In Vidarbha, it is not perfect K = 3 for number of towns of all classes. The value comes to 3.35 with the help of the following formula. (Sawant & Bhole, 1980, p. 831)

$$K = \frac{K1 + K2 + K3 --- Kn}{n}$$

$$K1 --- Kn = \frac{NL}{NH}$$

Where,

NL = number of towns of the lower order, and

NH = number of towns of the next higher order.

The K value more than 3 means the towns have to draw their clientele from area equivalent to more than 3 tributary areas. The goods and services in the town are sustainable with service areas equivalent to 3.35, in Vidarbha.

The distribution of towns with their hierarchical order is presented in Fig. 3.4. First of all, there stands out the randomness in the distribution of towns. The nearest neighbour statistic (Rn = 2D$\sqrt{N/\ }$) yields the value of 0.99 as applied to the point distribution, establishing that it is more random than uniform. Though there is no significant clustering, the deviation from regularity is unmistakable. The irregularity in the distribution of towns is bound to distort the network pattern.

The network is not only distorted (not fully shown in Fig. 3.4 as it will be too confusing) but it is not hexagonal also. The hexagon of the third order places around 2nd order Amravati is not complete in Vidarbha. So also, there are 4th order places around Amravati and the shape of the service area is a pentagon rather than a hexagon. It is the same with the shape formed by 5th order central places around Gadchiroli (Fig. 3.4). The hexagons are not formed by the central places in Vidarbha because they are formed according to Christaller, in a closely packed system. With an uneven distribution of population, Vidarbha does not have a closely packed system. So the market principal also works out to be more than K = 3.

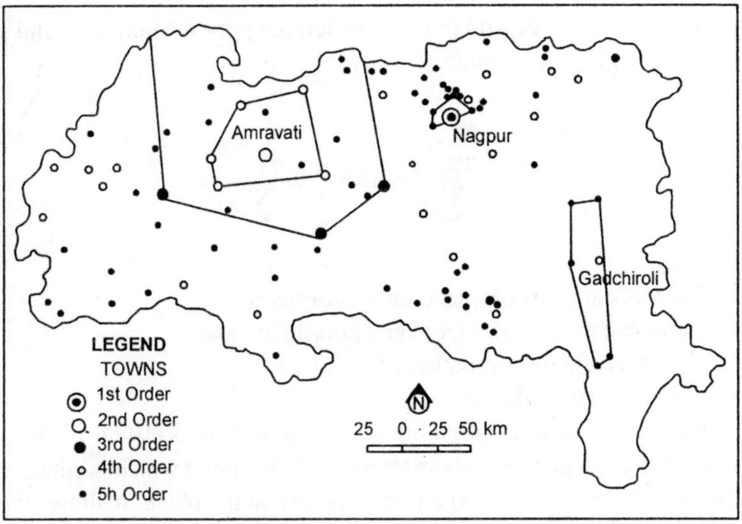

Fig. 3.4: Distribution of Towns in Vidarbha

Uneven distribution of population affects the size of the service areas as it affects their shape. The service area around Gadchiroli is much larger than that formed around Nagpur by the same 5th order central places (Fig. 3.4) Gadchiroli district has a very low population density of 55 persons per sq. km. while Nagpur district has high density of 332 persons per sq. km. Naturally Gadchiroli has to draw its clientele from a much larger area than Nagpur to sustain its central functions. Berry (1967, p. 90) also remarks that sizes at each level of hierarchy within a complex economy vary in response to changes in population density.

An empirical case like Vidarbha cannot come very close to Christaller's ideal model. However the central places are centrally located for the people. They strive as much as possible to be at a minimum distance for the maximum number of them. The service area of the city of Amravati bounded by 4th order central places is picked up as an example. The area is shown as enlarged and having point distribution of villages of 3000 + population within it in Fig. 3.5. The centrographic method (Fitzgerald, 1975, p. 33) is used to find out the gravity centre of population to decide whether Amravati is really located in the centre of the population. The gravity centre is the mean centre of population distribution averaging the total pull of population on the central place. It is derived with the help of x and y coordinates by the formula;

$$\overline{X} = \frac{\sum_{i=I}^{n} xiwi}{\sum wi} \text{ and } \overline{Y} = \frac{\sum_{i=I}^{n} yiwi}{\sum wi}$$

Where,

xi = distance of ith place on x coordinate

yi = distance of ith place on y coordinate and

wi = weight or population of ith place.

(Yeates, 1974, p. 31)

The method is bivariate but according to Yeates (1974, p. 30) it is directly analogous to univariate mean as that point which minimises the sum of the distances to all other points on the plane. It shows the central tendency.

The formula is applied to the service area of Amravati and the gravity centre is marked x. (Fig. 3.5). It is found to be just 10 kms away from the present location of Amravati. It does not exactly coincide with Amravati. But 10 kms distance may be taken as tolerable deviation for the location of a city and that of Amravati can be called spatially efficient.

Thus, there is a central tendency of the central places to be located at a minimum distance from maximum number of people in spite of varying size and shape of the service area. Such a tendency is favourable for spatial efficiency.

Spatial Efficiency in Location of PHC: A Simulation Model

Spatial efficiency model can be extended to simulate optimum locations for a facility. It means that those places are able to offer the facility to maximum number of people at a minimum average distance for them.

M.J. Hodgson and C. Valadares (1983, p. 49) have tried a simple P-median model in the problem of spatial efficiency of health centers in Salcette, Goa, by which the minimum of weighted distances to the facility is searched for. The model locates the facility in such a way that the distance to it for aggregate population is minimised.

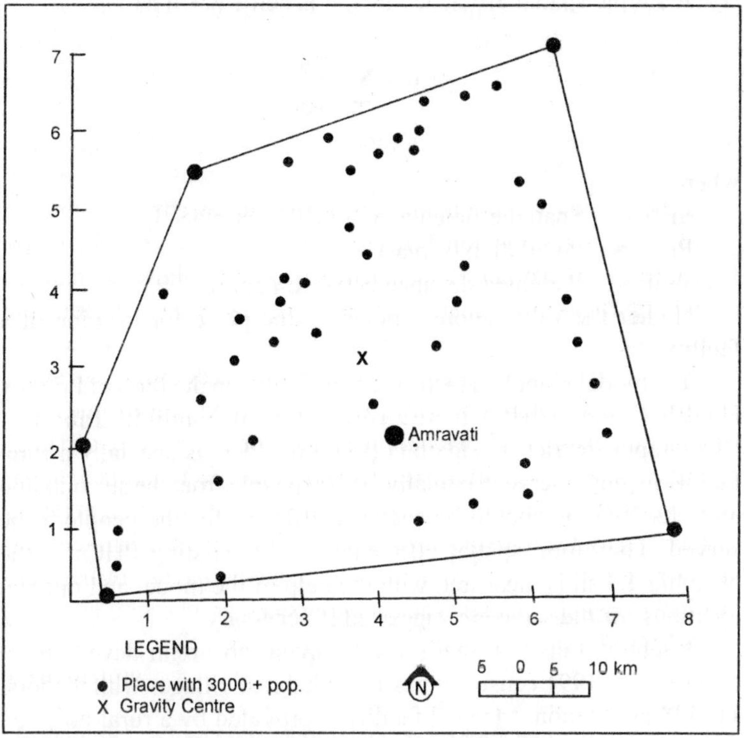

Fig. 3.5: Gravity Centre of Population

Bryan Massam and others (1984, pp. 66-75) used a formal linear programming model to test the results in the same region. Yeates (1963, pp. 7-10) also used a linear programming model to work out spatial efficiency of high school districts in Grant County, Wisconsin. J.F. Hart (1954, pp. 56-59) has applied graphical method for finding a modal point in settlement distribution which he calls a point of minimum aggregate travel.

All these methods put a stress on minimisation of distance. But spatial efficiency is affected by the pull of population as well which is a maximising factor. Therefore, the spatial efficiency model must include the maximisation of the pull of the people alongwith the minimum aggregate travel. Such a model is helpful not only in testing spatial efficiency of the existing location of the facility but also in simulating an optimum location for it. In this way, the model is normative and has an immense value in planning.

A simple spatial efficiency model is constructed as

$$SEfi = \sum_{\substack{i=1 \\ j=1}}^{n} \frac{Pi}{wdij}$$

where,

SEfi = Spatial efficiency of a facility at place i,
Pi = Population of place i,
Wdij = Weighted distance between i and j

Higher the value, more efficient is the place for location of a facility.

The model is applied to the medical facility in the form of Primary Health Centre (PHC) in the rural area of Nagbhid Tahsil of Chandrapur district in Vidarbha (Fig 3.6). PHC is specially a rural facility meant to serve maximally 30,000 people from the surrounding area. Its location should be spatially efficient for the people to be served. Therefore, spatial efficiency of the existing PHCs in the Nagbhid Tahsil is measured with the help of the model and optimal locations for the same are suggested if necessary.

Nagbhid Tahsil is totally rural, having no urban development according to 1991 census. It has 116 inhabited villages with the total 112728 population. Medical facility is provided by a rural hospital, maternity homes (2), family planning centers (2), dispensaries (7), child welfare centers (2) and private practitioners (23) besides 8 PHCs now located in village Sulezari, Nagbhid, Moushi, Govindpur, Bond, Balapur Budruk, Talodi and Wadhora. Out of these, Nagbhid (the tahsil place) and Talodi do not need an additional PHC when other medical facilities like hospital, maternity home, child welfare center, dispensary etc. are available.

These PHCs can be shifted to some other rural places having no medical facility, to serve the people better.

Constraints: There are some constraints on computation of spatial efficiency. They are:

(i) Villages with 200 + population are considered for computation since lower population will have little effect on location of PHC.

(ii) The villages with 1000 + population only are considered as prospective places for location of PHC.

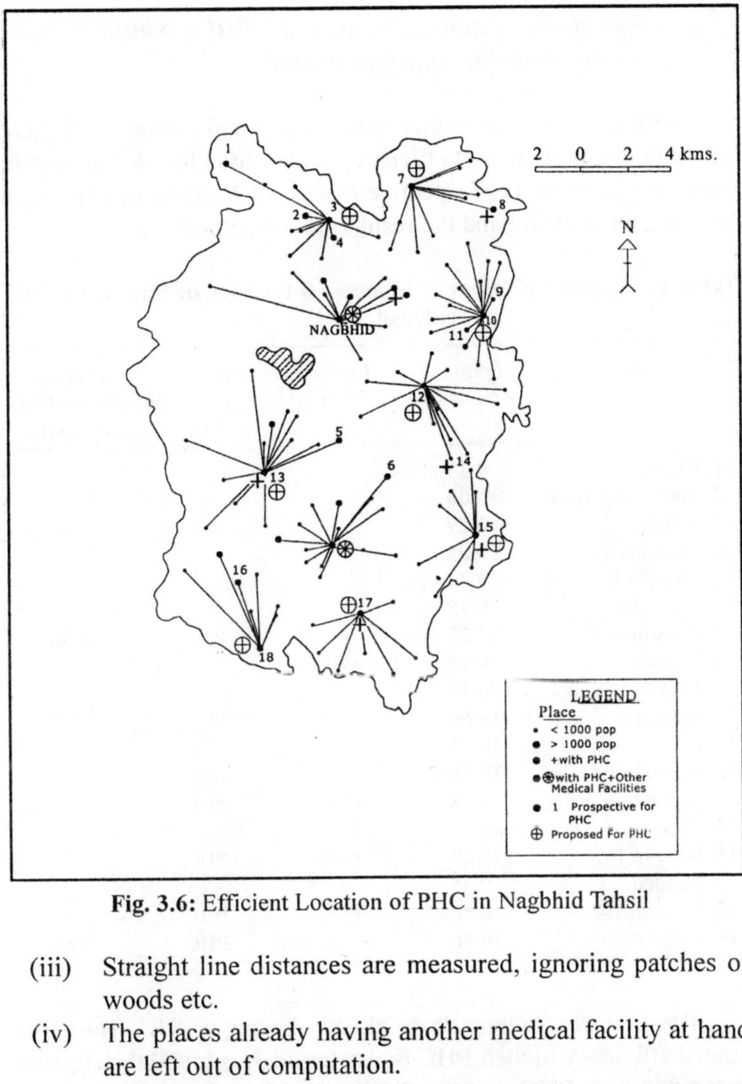

Fig. 3.6: Efficient Location of PHC in Nagbhid Tahsil

(iii) Straight line distances are measured, ignoring patches of woods etc.

(iv) The places already having another medical facility at hand are left out of computation.

(v) Villages within 5 kms of a rural hospital (*e.g.* at Nagbhid) and those within 2.5 kms of any other medical facility except a PHC are also left out, since they can avail of the facility within a short distance.

(vi) If two or more adjacent places within 5 kms from each other showed high efficiency values, the one with higher value is chosen as optimum, so that the PHCs would be well distributed in the rural countryside.

There are 8 PHCs existing in the region and they are sufficient for the population. Spatial efficiency is measured for 18 places with 1000 + population as prospective places for location of PHC with the help of the model and the results are tabulated below:

Table 3.2: Spatial Efficiency of Villages for Location of PHC in Nagbhid Tahsil, 1991.

Place	*Spatial Efficiency*	*Existing PHC*	*Chosen for Location of PHC*	*Average dist. to PHC in Kms.*
1. Khapa	9.35			
2. Mahadi Mokasa	10.06			
3. Mangli	10.29		PHC	3.38
4. Bhahamani	10.05			
5. Chindhi R.	9.37			
6. Janakpur	9.35			
7. Päharni	12.25		PHC	4.00
8. Moushi	9.45	PHC		
9. Kirmiti Mendha	10.15			
10. Kirmiti	11.60		PHC	3.13
11. Pardi	10.05			
12. Mindala	15.86		PHC	3.52
13. Govindpur	13.48	PHC	PHC	4.00
14. Bond	8.36	PHC		
15. Balapur Br.	10.09	PHC	PHC	3.71
16. Nanded	9.35			
17. Wadhona	10.50	PHC	PHC	3.78
18. Girgaon	10.46		PHC	4.64

These places include those already having a PHC. Therefore, spatial efficiency of their PHC is also tested. It is found that 3 places can retain their PHCs but the PHC at 2 places should be shifted to more efficient places along with 3 PHCs where other medical facilities exist. (Two places, Nagbhid and Talodi, themselves have the facilities and the third place is very near Nagbhid and can avail the facilities there) Fig. 3.6 shows the distribution of efficient PHCs. The average distance for the people to the proposed PHC is less than 5 kms.

Thus the model yields 8 efficient locations for the facility of PHC in the rural countryside of Nagbhid Tahsil. Tidswell and Barker (1971, pp. 23-26) have suggested a simulation grid to find the probability of location of a facility with running scores. The method is stochastic and has a chance element. The spatial efficiency model (SEfi) is simple but useful and gives satisfactory results. The same is true of the P-Median model of Hodgson & Valadares (1983) which uses weighted distances and finds spatial efficiency.

Conclusion

The central places have a central tendency to locate in their service area to minimise the distance for the clientele to travel. On this basis of minimum distance concept, Christaller produced his renowned central place theory which envisages the central locations of the central places in the most perfect and efficient hexagonal lattice of service areas. In real world cases, the network is much distorted in size and shape due to irregularities in the distribution of central places, their population and the economic conditions of the people. Still, the central places strive to be as much spatially efficient as possible by being at minimum distance for maximum number of people. This principle of spatial efficiency can be successfully used to simulate the location of any facility in a place for planning purposes. The operational model measures spatial efficiency of the existing facility and proposes its better location.

Efficiency in travelling is always sought by man everywhere as in case of central place location. The routes he uses for travel are important for the purpose. The next chapter turns to spatial efficiency of routes and route network, one of the most interesting growing points in Geography.

REFERENCES

Alonso W. (1964), Location Theory, in L. Needleman (ed) *Regional Analysis*, Penguin Books Ltd., Harmondsworth, England, p. 371.

Berry B.J.L. (1967): *Geography of Market Centres and Retail Distribution*, Prentice-Hill Inc., Englewood Cliffs, New Jersey, p 86.

Ibid, p. 90.

Carter H. (1972): *The Study of Urban Geography,* Edward Arnold, London, p. 69.
Ibid, p. 115.
Cole J.P. and C.A.M. King (1968): *Quantitative Geography,* John Wiley & Sons Ltd. London, p. 492.
Doxiadis C.A. (1969): Ekistics: *An Introduction to Science of Human Settlements,* Oxford University Press, New York, pp. 29-37.
Fitz Gerald B.P. (1974): *Developments in Geographical Methods,* Oxford University Press, London, p. 33.
Garner B.J. (1967): Model of Urban Geography and Settlement Location in R.J. Chorley and P. Haggett (eds.), *Socio-Economic Models in Geography,* Methuen & Co. Ltd., London, p. 304.
Ghosh S. (1998): *Introduction to Settlement Geography,* Orient Longman, Calcutta, p. 65.
Ibid, p. 64.
Haggett P. (1968): *Locational Analysis in Human Geography,* Edward Arnold, (Publishers) Ltd., London, p. 49.
Hart J.F. (1954): Central Tendency in Areal Distributions, *Economic Geography,* Vol. 30, pp. 56-59.
Hodgson M.J. and C.Valadares (1983): The Spatial Efficiency of Health Centers in Salcette, Goa, *Annals of the National Association of Geographers,* India, Vol.3 No.1, p. 49.
Mandal R.B. (1981): *Planned Development of Rural Settlements,* Concept Publishing Company, New Delhi, p. 373.
Massam B.H., C.P. Singh, A. Manahan & I. Askew (1984): The Location of Health Care Facilities in Salcette, Goa, *Annals of the National Association of Geographers,* India, Vol 4, No 2, New Delhi, pp. 66-75.
Sawant S.B. and A.S. Bhole (1980): Distribution, Centrality and Hierarchy of Central Places in the Indrayani Basin, *The Deccan Geographer,* Vol. XVIII, No. 2, Secunderabad, p. 831.
Tidswell M.V. and S.M. Barkar (1971): *Quantitative Methods, An Approach to Socio-economic Geography,* University Tutorial Press Ltd. London, pp. 23-26.
Toyne P. and P. Newby (1972): *Techniques in Human Geography,* McMillan Education Ltd., London, p. 105.
Yeates M. (1974): *An Introduction to Quantitative Analysis in Human Geography,* McGraw-Hill Book Company, New York, p. 31.
Ibid, p. 30.
Ibid, pp. 7-10.
The Maharasthra Census Directorate: District Census Handbook 1991 -Districts- Akola, Amravati, Bhandara, Buldhana, Chandrapur, Gadchiroli, Nagpur, Wardha and Yavatmal. The Census of India, Maharashtra State, Mumbai.

4

SPATIAL EFFICIENCY IN LOCATION OF ROUTES AND MOVEMENT

Location of routes and route networks is vital in spatial organisation of human activities as they provide linkages for movement. For the geographer, the importance of transport routes lies in its being one of the principal factors affecting the distribution of social and economic activities. According to Tolley and Turton (1995, p. 4) transport systems have a geographical dimension and relevance. They emphasise the vital interaction between transport and other essentially spatial processes such as industrial location and urbanisation.

Routes are more or less fixed channels of traffic flows between source and destination points and the connected routes form the network. The networks of roads, railways, airlines, pipelines, cable lines etc. carry goods, peoples, information and messages. The quantitative revolution of 1960s provided transport geographers with valuable methods to investigate these routes and networks with greater precision. The work of Kansky (1963), Warntz (1966), Nystuen and Dacey (1961), Taaffe, Marrill and Gould (1963), Garrison and Marble (1974) have given a new dimension to the study of route networks. They are now more and more analysed with the help of graph theory which happens to be the part of topology in Mathematics. Any transport network may be considered as a topologic graph for quantitative measurements with the help of its two basic characteristics among its different structural properties *viz.* nodes and lines or vertices and edges. The circuity, connectivity and accessibility of transport network can easily be subjected to topological measurement.

It is also in this field of transportation that modern technology has virtually revolutionary effect. It has made variety of modes of transport and communication possible and has increased the network by leaps and bounds. Technological improvement is being utmost felt in flows of information through communication system. Satellites with great capacity for transmitting information and fibre optic networks have proved their efficiency on the national as well as international scale. The information technology through computer networks has radically improved the communication techniques. These technologies have also enabled the transport to operate more efficiently and more reliably. However, this chapter is confined to the study of transport routes and networks and movement on them. Lines of communication are not taken into account as they are little concerned with space.

Spatial efficiency can be investigated in the location of routes and movement on them as its two factors—distance minimisation and maximisation of activity apply there.

Location of Routes

There are several factors, influencing the location of routes—the historical, technological, spatial, economic and political factors. The route network always has a historical legacy as it develops through time. In India, the trunk railways were laid by the British during the British Raj which are still serving the people. The route networks and their prosperity are undergoing fast changes, especially in modern times, due to the transport technology. Janaki (1985, p. 71) remarks that modern transport has changed the whole concept of space relationship on the surface of the earth. The world has been ushered in an era oi speed and comfort never known before. Car ownership has steadily expanded to eclipse the railways in the developed countries. In the UK, the canals like the railways were unable to match the advantages of road haulage for industrial traffic and after 1920, they rapidly declined as the commercial routes. (Tolley and Turton, 1995, p. 35). The channel tunnel built underwater between the UK and France is an excellent example of how technology can work wonders in route building. The piggyback (Trailor-on-flatcar) or fishy-back (container-on-flatcar) operation is another example of tremendous changes taking place in transportation.

However, the spatio-economic factor rules supreme in the location and construction of route network which must be economically viable and spatially suitable. It must be able to generate the traffic to offset the costs. Capital is invested in the construction of routes perceiving the benefits in terms of revenue and the potential for economic growth in the region even in welfare projects. In 1869, the Suez Canal and in 1914, the Panama Canal were expensive to build, but their prospects for the shipping made them a worthwhile investment. However, maritime trade through these canals has been affected in the last 50 years due to their inability to admit large vessels, remarkable for the trade of the modern times. At the same time, distance is also involved alongwith economic considerations in the location of routes. To save distance so as to save friction of space and effort in it is always a prime motive. That is why the routes are kept as short as possible. The Suez and the Panama Canals rapidly rose to prominence because they cut on distance quite a great deal. In case of land routes, their relationship with the physical landscape cannot be over emphasised. Geographers have studied the constraints of relief on the land routes and how they avoid the high grade to minimise the building and operating costs.

The routes and their networks are often discussed in terms of the builder's cost and the user's cost. The minimisation of the former increases the later and vice versa and a compromise between the two is desirable.

Movement on the Routes

Routes are created for the movement of people and goods and movement is inseparable from the routes. Movement necessarily involves friction of space in overcoming distance and it has always been an attempt of man to reduce the friction or effort in movement. Movement minimisation has therefore, been a basic concept in transportation and not the volume of movement as such. It means minimisation of time/cost of transportation. So, man has contemplated movement minimisation in two ways (a) by increasing speed and (b) by reducing distances.

Man is increasing speed in traversing distances in order to minimise the barrier effect of distance and he has achieved such a great success in it that the world is said to have virtually shrunk. In the 21st century, it will be a matter of little time to reach any corner of the world. In the 20th century itself, highly efficient mechanised transport began to affect the global economy through its influence on manufacturing, agricultural patterns and urban expansion. The comparatively lower levels of economic activity in the developing countries are often attributed to the modest scale of their transport system.

Movement minimisation is most effectively achieved by minimising distance itself and it leads to efficiency. Shorter the distance, better it is for efficiency in movement. To cut the distance short is a general tendency of man even in everyday life. The straight route has the shortest distance and airlines and oceanic shipping lines where the routes are not built follow the straight course as far as possible.

For movement minimisation, location of routes is very important. Routes should be laid in such a way that distance is minimised and so is the movement along the route. For the builder and user of the route, it is the minimisation of cost. Longer movements tend to be more costly and this is important given the tendency of the people to minimise cost. (Cox, 1972, p. 21). However, movement minimisation does not at all mean reduction in traffic flow. On the other hand, movement is expected to increase in volume so that revenue increases. The location of routes play an important role in this respect also. They are desired to go the way that procures more and more traffic.

Spatial Efficiency in Location of Routes and Networks

Spatial efficiency in the location of routes and their networks is also judged on the basis of its two factors: (i) minimization of distance or route length and (ii) maximization of activity or traffic flow. Gauthier (1974, p. 172) also opines that efficient movement means maximum possible movement with minimum possible effort in travel. Actually the optimum lies in a compromise between the two which happens to be an efficient solution. The route length may be somewhat increased above the minimum if it yields more traffic, or some traffic may be ignored if it increases the route length too much. Wellington's (1886) pioneer work in railway construction between Mexico city and Vera Cruz in Mexico proves the point. He was an engineer faced with the problem of planning the railway (Haggett, 1968, p. 62) and he finally reached the optimum solution which lay neither in minimum route length (straight-line) nor in maximum traffic (connecting all the places) but in compromise between the two. The efficient solution called for somewhat increase in route length above the minimum and some reduction in maximum traffic by leaving out some places. It is also a compromise between the builder's and the user's costs. So, spatial efficiency should be measured and simulated with respect to both the maximum-minimum factors. Many times, the absolute efficiency may not be reached but it has to be maintained, at least, on satisfactory level.

Circuitry and connectivity, accessibility, and detouring are major locational aspects of routes and their networks. They are discussed with the spatial efficiency point of view.

Circuitry and Connectivity

Interaction between places requires connecting links which produce connectivity of the route network between a set of origins and destinations, an indispensable part of any economy. More the connections, better is the connectivity. Circuitry denotes a closed path (Garrison and Marble, 1974, p. 61) and a well-connected network is often circuitous. Connectivity and circuitry may be taken as complementary and co-varying terms.

Networks are now more and more analysed with the help of graph theory, using its basic structural properties of nodes and links. Any transport network may be now considered as a topologic graph for quantitative measurements (Hodder and Lee, 1974, p. 156). There are four structural indices namely, the cyclomatic number (μ), the alpha index (α), the beta index (β) and the gamma index (γ), which provide a measure of quantum of connectivity and thus the levels of nodal accessibility. Higher the values, better is the circuitry and connectivity in the region.

With better connectivity, the possibility of direct and shortest routes between pairs of nodes increases. But it does not mean that spatial efficiency increases with it. That is why maximum connectivity may not be maximally spatially efficient. On the other hand, it may render the network spatially inefficient. For, spatial efficiency depends not only on the shortest routes but also on the maximum service of them. As the number of shortest routes or number of one-route connections increases, their use decreases and efficiency is adversely affected. The same thing may be expressed in terms of cost-to-the-user and cost-to-the-builder (Lloyd and Dicken, 1972, p. 74). A completely connected graph is ideal for the user as he has to pay the least for the shortest routes but not at all for the builder who has to pay high sums for paths more than necessary for service. Contrarily, the minimization of construction cost increases the user's cost and the optimum lies between the two. This is well illustrated by six different networks of Bunge (Haggett, 1968, p. 65) (Fig. 4.1) in which neither the poorly connected graph (A), nor the completely connected graph (D) is ideal. In other words, neither can be spatially efficient. It is the one (C) in which both the costs or the total route length and the number of direct and short connections are well balanced. The direct or one-route connections minimize the distance and the route utility increases by restricting the total route length. This is achieved if same key nodes are used more intensively than others. In reality also, it is a common experience that there are a few nodes in a network whereupon many routes converge, thereby creating a number of one-route connections but at the same time, all the nodes are not provided with direct connections, thereby restricting the total route length. Thus, spatial efficiency is enhanced not by only high degree of connectivity disregarding the route utility but by a proper compromise between them.

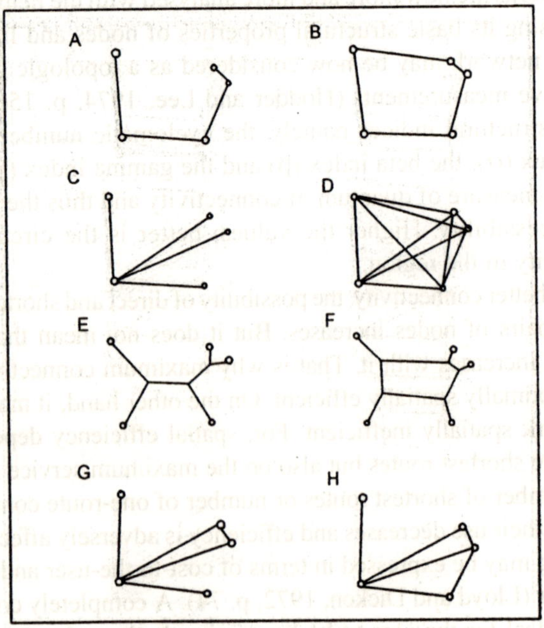

Fig. 4.1: Route Network Graphs

Spatial efficiency of connectivity can be measured by the following formula in graph theoretic language

$$SEC = \frac{L/e}{Rx1 + Rx2 + Rx3 + \ldots + Rxn}$$

Where,

SEC = Spatial efficiency of connectivity;
L = Maximum connectivity;
e = edges;
Rx1 = One-route connection,
Rxn = n-route connection.

Maximum connectivity (L) is derived after Prihar (1956, pp. 929-933) using the equation

$$L = \frac{V(V-1)}{2}$$

Where, V = no. of vertices in the network.

With the above formula of SEC, relative efficiency of connectivity is usefully measured and more efficient network is found out. Higher the value, better is the spatial efficiency and the highest value is the optimum. Spatial efficiency is a delicate balance between route utility and total no. of n-route connections by the shortest path. Route utility decreases with increasing number of links or edges. It can be derived if the total route length in the network is compared against maximum possible connections in it. So also, one-route connections are the shortest paths. But with increasing number of two or more route connections, distance increases and efficiency is adversely affected. With the above formula of SEC, spatial efficiency of Bunge's six networks (Fig. 4.1) has been measured and the results are tabulated below. To measure relative efficiency two more graphs are produced with more edges simulated and results are given in the same Table.

Table 4.1: Spatial Efficiency of Route Network Graphs

Graph	*L*	*Edges*	*Rx1– + Rxn*	*SEC*
A	10	4	20	0.125
B	10	5	16	0.125
C	10	4	16	0.156
D	10	10	10	0.100
E	28	7	45	0.089
F	45	10	53	0.085
G	10	5	15	0.133
H	10	6	12	0.139

In the above example the graph C has the maximum value and it is, therefore, the most efficient network. Graph D is maximally connected but since its total route length is too much, efficiency decreases. Graph C has minimum route length but distance between the nodes is short, only upto 2-route connections. Therefore the efficiency is utmost. In graph G, one connection and in graph H, two connections are added to graph C network for simulation which increases the no. of one-route connections. But route utility decreases varyingly with increased route length and spatial efficiency is reduced, but it is still more than that in completely connected graph.

Graph C is also ideal from the point of view of the builder's cost and the user's cost, as both are well balanced. In maximally connected graph D, the builder's cost is unnecessarily high while in the poorly connected graph A, user's cost soars high. The best way, thus, to achieve network efficiency or the cost balance is to increase the nodality of some key nodes in the network by incidence of routes upon them.

Spatial efficiency of connectivity of any existing network can usefully be measured with the formula of SEC and it can also be simulated by adding new edges or connections for planning purposes. Building and operating costs may also be taken into account. Prihar (Garrison, 1974, p. 86) points out different costs in designing networks. In case of maximum route efficiency pursued by geographers, Tolley and Turton (1995, p. 25) opine that the solution of the shortest-path problem lies between the two extremes of complete interlinkage and a simple form of circuitous routing as a compromise where building and operating cost constraints are combined and balanced.

Accessibility

Connectivity of the network and accessibility are closely related. A well-connected network leads to good accessibility since the places are easily reached by routes. The Independent Commission on Transport (1974) concluded that the real meaning of mobility or the true goal of transport is access. The spatial relationship between origin (supply) and destination (demand) in the form of links is an essential component of accessibility and a measure of the extent to which demand for transport is realised.

Ideally, each place should be easily accessible to transport route, at best within 5 kms from it. Beyond it, accessibility decreases progressively. This can be mapped by distance isopleths of L.D. Stamp (Monkhouse and Wilkinson, 1976, p. 280) which illustrate the proportion of inaccessible to accessible areas (Fig. 4.2). Tolley and Turton (1995, p. 21) have shown access by bus to hospitals with journey time contours in South Wales, UK. The route network is generally laid in such a way as to first connect the big industrial centres and urban places to each other and then to smaller places in rural areas, while a large part of the rural country-side remains inaccessible. It is because the routes follow high density areas to serve maximum people and to procure maximum traffic. As Warntz (1966, p. 61) pointed out, major transportation arteries in the USA occur on major ridge lines of potential population surfaces. In their analysis of Turkish railroads, Kolars and Malin (1974, p. 112) have also shown the correlation between the railroads and population peaks and ridges.

Accessibility of a place on the route or nodal accessibility according to Morrill (Akpan, 1990, p. 283), is the relative degree of ease with which it may be reached from other places. In a way it is measured on the basis of its associated number, better known as the konig number which is the topological distance in terms of number of edges from a place to the farthest point by the shortest path. Lesser the konig number, better is the accessibility of a place, and the most central place has the least konig number. Similar kind of accessibility measure is suggested by Shimbel which Garrison (1974, p. 89) used for the Interstate Highways in the SEn USA. Nodal accessibility has been analysed by P.A. Akpan in the eastern states of Nigeria with the formula—

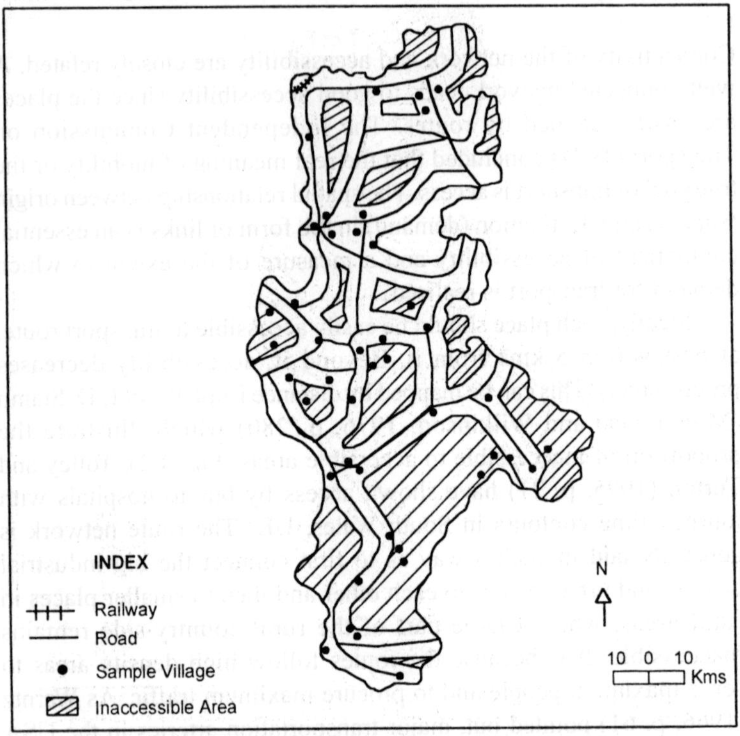

Fig. 4.2: Road Accessibility of Places in Gadchiroli District

$$Ai=\sum_{j=1}^{n} ai$$

Where,

Ai = Accessibility index for node i, aij = the shortest path distance between nodes i and j (Akpan, 1990, pp. 283-284). The smaller the index, the more accessible is the node and hence the higher its rank order.

However, spatial efficiency in accessibility is not determined by the accessibility index alone. It is the minimum factor of spatial efficiency but it has to be related to the maximum factor, in the form of traffic or the population served. If so related, it may be found that the place having the least accessibility index is not the most efficient one, or the vice versa.

Spatial efficiency of some key nodes on the network of airways of India (Fig. 4.3) has been measured with the help of their accessibility index as suggested by Akpan and number of incident routes on the nodes, which can be a good indicator of traffic or air service. More the incidence of routes, better it is for spatial efficiency. The simple formula to compute it is as follows:

$$SEai = \frac{IRi}{Ai}$$

Where,

SEai = Spatial efficiency of accessibility of place i, IRi = No. of incident routes on place i, and Ai = Accessibility Index of ith place.

Accessibility indices of the key nodes of Indian air-ways are derived by the formula given by Akpan and are produced in a matrix below.

This accessibility index of each selected node is used as a denominator which is a minimising factor related to number of incident routes of the respective nodes which is a maximizing factor in the computation of spatial efficiency. Higher the value, more is the spatial efficiency. The results are given in the following Table along with rank order according to accessibility index and spatial efficiency for comparison.

It is very clear from the Table 4.3, that the rank orders according to accessibility index and spatial efficiency do not correspond except in case of Mumbai and Delhi which means that lower accessibility index does not always indicate higher efficiency unless it is coupled with better traffic conditions. A good example is provided by Nagpur which is situated in the centre of India, has fair accessibility with 4th rank order but low spatial efficiency as the incident routes or traffic is low. Therefore, only the accessibility is not sufficient to analyse the efficiency of routes, although it is very helpful to know the spatial efficiency.

Detouring

The shortest route happens to be the least effort route and preferred by the people. The shortest route is the straight one. That is why the sea routes and air ways take a great circle course as far as possible to be straight and short on the spherical earth.

Table 4.2: Accessibility Matrix of Indian Airways

Nodes	*1.Ahm*	*2.Ban*	*3.Cal*	*4.Che*	*5.Del*	*6.Guw*	*7.Hyd*	*8.Koc*	*9.Mum*	*10.Nag*	*11.Sri*	*Acc. Index*
1. Ahmadabad	–	2	2	2	1	3	2	3	1	2	2	20
2. Bangalore	2	–	2	1	2	3	1	1	1	2	3	18
3. Calcutta	2	2	–	1	1	1	1	2	1	1	2	14
4. Chennai	2	1	1	–	1	2	1	1	1	1	2	13
5. Delhi	2	2	1	1	–	2	1	2	1	1	1	14
6. Guwahati	3	3	1	2	2	–	2	3	2	2	3	23
7. Hyderabad	2	1	1	1	1	2	–	2	1	2	2	15
8. Kochi	3	1	2	1	2	3	2	–	2	2	3	25
9. Mumbai	1	1	1	1	1	2	1	2	–	1	2	13
10. Nagpur	2	2	1	1	1	2	2	2	1	–	2	16
11. Srinagar	2	3	2	2	1	3	2	3	2	2	–	22
Acc. Index	20	18	14	13	14	23	15	21	13	16	22	

Table 4.3: Spatial Efficiency of Key Nodes on Indian Airways

Nodes	*Acc. Ind*	*IR*	*SEai*	*Rank order (Acc. Ind)*	*Rank order (Spatial Efficiency)*
1. Ahmadabad	20	4	0.20	6	9
2. Bangalore	18	6	0.33	5	6
3. Calcutta	14	11	0.78	2	3
4. Chennai	13	9	0.69	1	4
5. Delhi	14	13	1.00	2	2
6. Guwahati	23	5	0.22	9	8
7. Hyderabad	15	7	0.47	3	5
8. Kochi	21	2	0.10	7	10
9. Mumbai	13	15	1.15	1	1
10. Nagpur	16	4	0.25	4	7
11. Srinagar	22	2	0.09	8	11

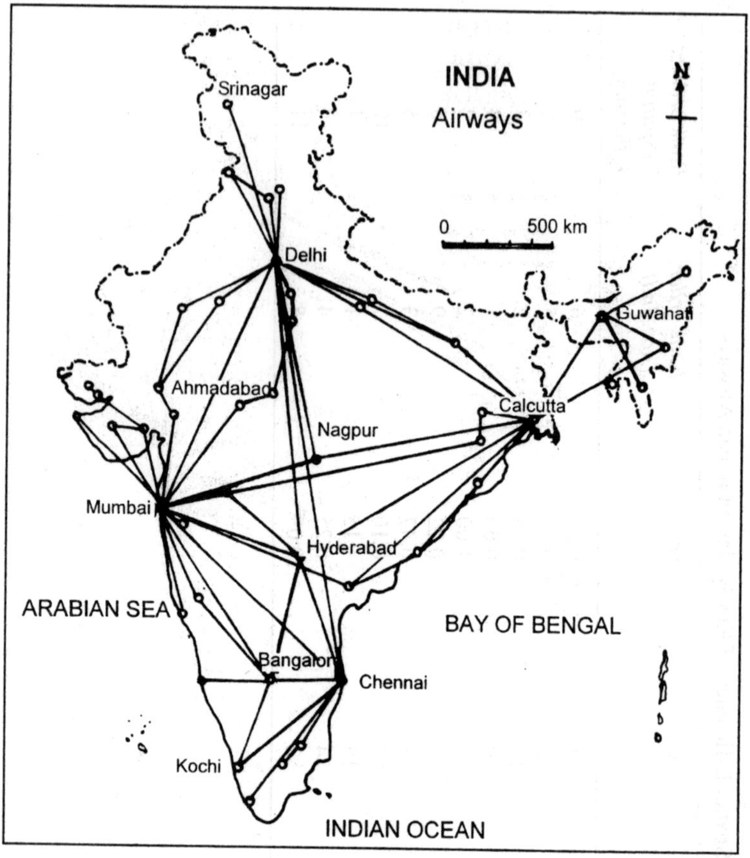

Fig. 4.3

However, the land routes are hardly straight due to various reasons. They are bound to detour with bends and curves. The amount of detouring can be quantitatively expressed by the detour index which is a simple ratio between the actual route distance and straight-line or direct distance between two places in per cent.

$$\text{Detour Index} = \frac{\text{Actual Distance}}{\text{Straight} - \text{Line Distance}} \times 100$$

Less the ratio, less is detouring and straighter is the route. Thus, the detour index measures the deflection of route from the straight course.

Detouring is necessary when either there are barriers to avoid-physical, economic, political etc. or there are important places to link up. Losch commented on the application of the laws of refraction to the study of route deflection. He suggested that the mountain barrier bends the route just as an optical lens would refract a beam of light: higher the refractive index, greater will be the amount of deflection (Lloyd and Dicken, 1972, p. 76), Warntz (1961) also states that even airline routes are not simple straight-line links between origin and destination because of pressure and wind effects.

The most common barrier to a land route is physical in the form of a mountain, a water body or a forest which the route has to circumvent by detouring. Railway construction in Bolivia was severely restricted by the Andean mountains with heights over 4000 m and by the poorly drained oriented lowlands. The maritime freight traffic is easier between the eastern and the western sea coasts of the USA through the Panama Canal rather than the inland route across the Rockies, though the former requires detouring.

Many times, detouring is spatially efficient even though it increases the route length or distance to be travelled. It is a common knowledge that the paths in the hilly areas are spiral to avoid high grade. Such a detouring path is spatially efficient because the shortest and straight path up the hill is more orduous, demanding more effort than the longer winding path.

Detouring is also desirable and efficient if it increases the utility of the route so as to put off the disadvantage of the increased route length. For, spatial efficiency depends as much on the utility of the route (greater the better) as on the route length (shorter the better). It is a compromise between the two. Fig. 4.4 shows 3 routes with their lengths and traffic. Spatial efficiency for them works out to be 0.45, 0.60 and 0.64 for A, B and C respectively. Obviously, the efficient route is neither A nor B but C, a compromising route between length and traffic.

Wellington, an engineer, constructing railway in Mexico, referred earlier, faced the same problem whether to keep the railway straight or to detour it, joining all the places between Mexico city and Vera Cruz. The solution he arrived at, to join only important urban and industrial centres by somewhat detouring the railway line for the purpose, was really spatially efficient, as the detouring was expected to yield traffic more than enough to offset the increased railway length.

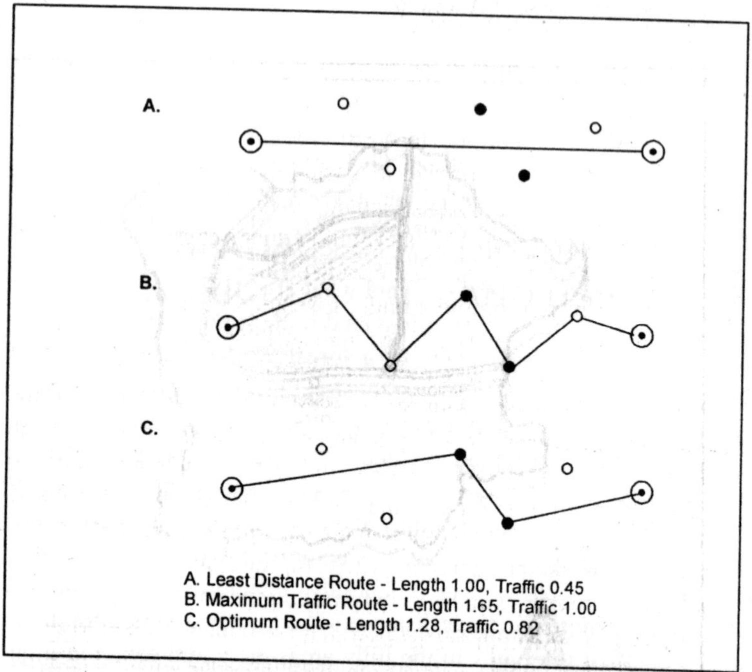

Fig. 4.4: Optimum Route

Black's (1993) discussion of route location points to the same principle of spatial efficiency in detouring which he calls a theoretical economic approach as illustrated by Tolley and Turton (1995, p. 49). Three railways in the US state of Indiana, radiating from Indianapolis to (a) Peru, (b) Madison and (c) Lawrenceburg are aligned on a theoretical basis so as to identify the least cost routes. The alignments are detoured in order to maximize the traffic as far as possible while keeping the route length as short as possible. This unequivocally indicates an attempt at spatial efficiency. In all these three cases, strong similarities between the actual and generated alignments have been observed, clearly showing the general tendency of man to be spatially efficient.

Spatial efficiency in detouring can be quantitatively measured by two parameters namely, traffic or population served en route and the detouring of the route as derived by the detour index. The formula is:—

$$SEd = \frac{\sum_{i=1}^{n} Pi}{DI}$$

where

SEd = Spatial efficiency in detouring.
Pi = Population of ith place en route.
DI = Detour Index of the route.

Higher the value, better is spatial efficiency.

The measure is applied to the detouring roads between Bhandara and Gondia in Bhandara district of Vidarbha (Fig. 4.5) where there is no straight route due to dense forest. The results are tabulated below:

Table 4.4: Spatial Efficiency of Road Detours in Bhandara dist.

	Road	*Population*	*DI*	*Spatial Efficiency*
1.	Bhandara -Sakoli-Gondia	66673	160.0	416.7
2.	Bhandara-Tumsar-Gondia	240311	144.0	1668.8
3.	Bhandara-Tirora-Gondia	201792	132.0	1528.7

The longest route by detour is through Sakoli (1) and it is not spatially efficient as it is also not serving much population. The route through Tumsar (2), serves highest population and is more efficient though longer than the one through Tirora alone (3). Moreover, the road from Tirora to Bhandara is unmetalled and so less desirable. (Fig. 4.5)

People have a tendency to choose spatially efficient route. However, as Gauthier (1974, p. 173) puts it, this does not mean the transport user is an optimizer. His decision is made within a range of choices having definite economic limits. Spatial efficiency consideration is able to help him in his choice.

Simulation of Spatial Efficiency in Bus Routes of a Town

Geographers are dealing more and more with the problem of movement minimization or finding minimum distance or the least-cost path. But it demands at the same time, maximization of flow. Thus, the problem of transportation requires the primal-dual solution. According to Cox

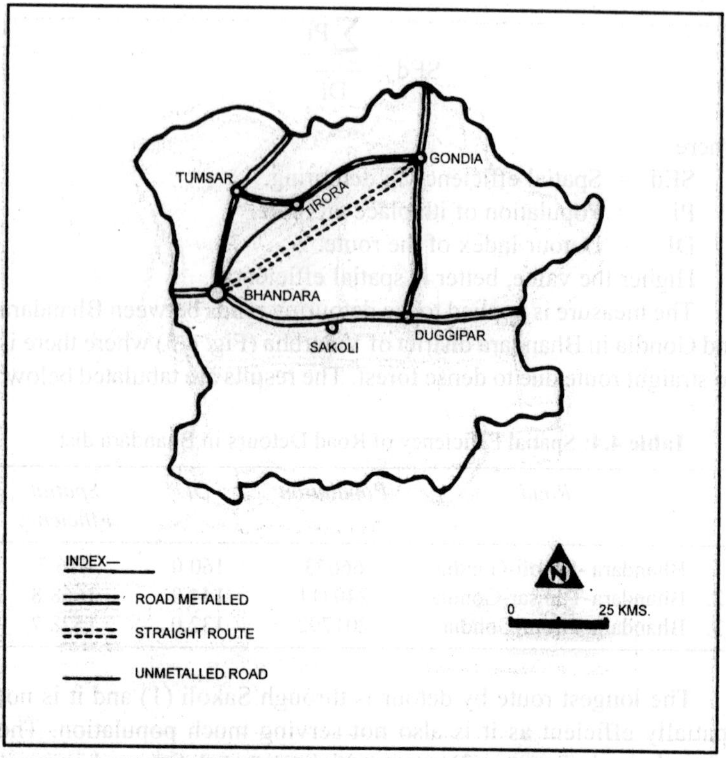

Fig. 4.5: Road Link in Bhandara District

(1974, p. 151), the solution to the transportation problem should provide us with something more than the minimal transportation cost which is also equal to a maximization problem. If we regard minimization with, maximization problem, the solution gives nothing but spatial efficiency.

A spatial efficiency solution has been sought in a problem of bus transportation in a hypothetical town from bus stops to down town area (Fig. 4.6). There are 20 stops and the maximum capacity of each bus is to take 60 passengers. The objective is to find spatially efficient movement path for 4 buses running in 4 parts of the town.

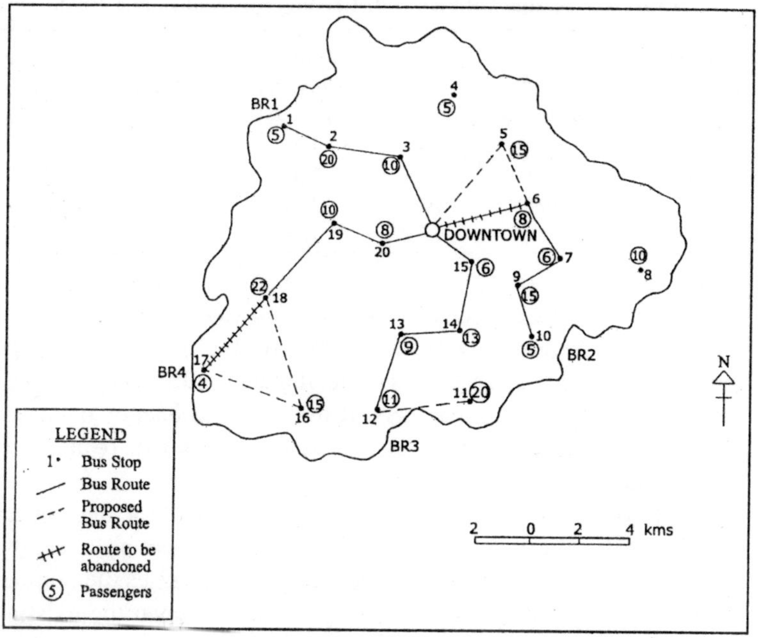

Fig. 4.6: Hypothetical Bus Routes in A Town

The Table 4.5 shows spatial efficiency of the existing bus routes at peak hour. There are some stops left out by the buses but since their capacity is not full, they can take more stops. The average passengers at peak hour at each stop are shown in the circle (Fig. 4.6) the problem is to add bus stop to a nearby bus route and to find spatial efficiency, which is computed with the formula,

$$SEbri = \frac{\sum_{s=1}^{n} Ps}{dbri}$$

where

SEbri = Spatial Efficiency of ith bus route
Ps = Average passengers of sth bus stop
dbri = distance or total length of ith bus route

Table 4.5: Spatial Efficiency of Existing Bus Routes in a Town

BR	$\sum_{S=1}^{n} Ps$	*dbri (kms)*	*Spatial Efficiency*	*Bus Capacity*	*Remaining capacity*
1	38	8	4.75	60	22
2	41	11.8	3.47	60	19
3	39	10.6	3.68	60	21
4	44	12.2	3.61	60	16

Since the bus capacity is not fully utilised, maximisation of average passengers is possible. Therefore, the left-out bus stops may be attached to a nearby bus route. It increases the route distance as well as number of passengers. If the spatial efficiency increases alongwith it, the attachment of stop is recommended. If not, it is preferable to be left out.

Additional bus stops are simulated for all the 4 bus routes following the North-West corner rule. For BR1, there are 2 nearby stop (Nos. 4 & 5) to be added. The addition maximises the number of passengers. But spatial efficiency decreases down the existing figure. The same thing happens if only one stop (No.5) is added and therefore no addition is recommended even though there remains much capacity to be utilised. Stop No.5 can be attached to BR2 which increases its spatial efficiency, (see Table 4.6).

Table 4.6: Simulation of Efficient Bus Routes in a Town

BR	*Stop no. added*	$\sum_{S=1}^{n} Ps$	*dbri (kms)*	*Spatial Efficiency*	*Remaining capacity*
1	+4, +5	58	16.4	3.53	2
1	+5	53	13.2	4.01	7
2	+5	56	15.4	3.64	4
3	+11	59	13.8	4.27	1
4	+16	59	15.6	3.78	1

while its capacity is nearly fully used. Now, since it cannot include any more stop, stop No.8 has to be left out and stop No.11 is attached to BR.3. Spatial efficiency of BR3 is enhanced by this addition within its capacity. But it cannot take stop No.16 which is attached to BR4.

This addition is recommended since it improves spatial efficiency of the routes.

Spatial efficiency model proves to be useful in such primal-dual problem of movement in various situations. Simulations are also possible and it helps in planning. Linear programming models are widely used in the transportation problem by geographers and as Yeates (1974, p. 243) has pointed out, the north west corner rule is a simple method to solve it. However, these models mostly deal with the primal problem of movement minimisation, while the maximising factor is held constant. If this factor is also considered in the linear programs, the solution will lead to spatial efficiency.

Conclusion

The route network is an infrastructure for the development of any region. But it is, in itself, a part of the development process and requires to be efficiently laid out. Three aspects of the network namely, connectivity, accessibility and detouring and their efficiency have been assessed in this chapter. So also, efficiency of the routes for movement is taken into account. Simulation of such efficiency is crucial in planning.

Migration is one of the important movements of mankind and it has its specific pattern, except in extraordinary conditions such as earthquakes, anarchy, war etc. Haggerstrand (1957) studied migration from Asby (Sweden) in the light of diffusion and found a strong distance-decay element. Stouffer (1940) produced an intervening opportunity model and argued that the opportunities were more important than mere geographical distance in migration. Hagerstrand also realised it and in his feedback model, he considered opportunities by the name of vacancies (Haggett, 1968, p. 47). Migration is thus, not only a function of distance but also of opportunities at destinations or at intermediate places. In other words, the analysis of migration in the light of maximisation of opportunities and minimisation of distance follows the principle of spatial efficiency.

Movement along the routes is a basic thing in the location of any economic activity. The location of industry is very much concerned with the distance factor in movement and spatial efficiency which the next chapter deals with.

REFERENCES

Akpan P.A. (1990), Changing Nodal Accessibility in the Eastern States of Nigeria, *The National Geographical Journal of India,* Vol. 36, Part 4, Varanasi, India, p. 283.

Ibid, pp. 283-84.

Cox K.P. (1972), *Man, Location and Behaviour; An Introduction to Human Geography,* John Wiley and Sons Inc., New York, p. 213.

Ibid, p. 151.

Garrison W.L. and D.F. Marble (1974), Graph Theoretic Concepts, in M.E.E. Hurst (ed), *Transportation Geography,* McGraw Hill Book Company, New York, p. 61.

Garrison W.L. (1974), Connectivity of the Interstate Highway System, in M.E.E. Hurst (ed.), *Transportation Geography,* McGraw Hill Book Company, New York, p. 86.

Ibid; p. 89.

Gauthier H.L. (1974): Least cost Flows in a Capacitated Network: A Brazilian Example in M.E.E. Hurst (ed) *Transportation Geography,* McGraw Hill Book Company, New York, p. 172.

Ibid, p. 173.

Haggett P. (1968), *Locational Analysis in Human Geography,* Edward Arnold (P), Ltd., London, p. 62.

Ibid, p. 65.

Ibid, p. 47.

Hodder B.W. and R. Lee (1974), *Economic Geography,* St. Martin's Press, New York, p. 156.

Janaki V.A. (1985), *Economic Geography,* Concept Publishing Company, New York, p. 71.

Kolars J. and H.J. Malin (1974), Population and Accessibility; An Analysis of Turkish Railroads, in M.E.E. Hurst (ed), *Transportation Geography,* McGraw Hill Book Company, New York, p. 112.

Lyod P. and P. Dicken (1972), *Location in space: A theoratical Approach to Economic Geography,* Harper & Row Publishers, New York, p. 74.

Monkhouse F.J. and H.R. Wilkinson (1976), *Maps and Diagrams,* Methuen and Co. Ltd., London p. 280.

Prihar Z (1956), Topological Properties of Telecommunication Networks, *Proceedings,* Institute of Radio Engineers, 44, p. 929.

Tolley R. and B. Turton (1995), *Transport System, Policy and Planning, A Geographical Approach,* Longman, Harlow, p. 4.

Ibid, p. 34.

Ibid, p. 21.

Ibid, p. 49.

Warntz W. (1966): The Topology of a Socio-economic Terrain and Spatial Flows, *Papers of the Regional Science Association,* Vol.17, p. 61.

Yeates M. (1974), *An Introduction to Quantitative Analysis in Human Geography,* McGraw Hill Book Company, New York, p. 243.

5

SPATIAL EFFICIENCY IN LOCATION OF INDUSTRY

We cannot think of modern economy without industries and their locational structure forms an indispensable part of spatial organisation of human activity. Therefore, the location of industries interest the economists and geographers alike (Chisholm, 1968, p. 2). Of course, the industrial production is an economic activity but its location implies spatial relations. The approach to industrial activity of a geographer is to understand the conditions that affect its location (Estall and Buchanan, 1968, p. 15).

The major contributions to location theory have been made by economists, mostly with a view to integrating location studies with the main body of economic theory. (McCarty and Lindberg, 1966, p. 182). They were concerned in their location studies with efficiency and profit-making. In every manufacturing plant, an important factor bearing upon efficiency is location. A well equipped and properly managed plant may fail in a poor location (Brisco, 1914, p. 53), and location is closely related to the distance which spatially separates the manufacturing plant from materials and markets, involving transport to overcome it. Distance, thus, lies between the demand and supply points. Therefore, distance in terms of transportation cost and/or time of assembling materials and delivering goods is an extremely important consideration affecting the location of manufacturing industry. If the movement costs of assembly or input are compared with those of distribution or output, the approximate location of industry would be nearer the higher cost point to save more costly of the two hauls. In a very simple two point problem of one source and one market, the orientation index (v) of industry can be derived as:

V = Qa/Qb where Qa is the assembly cost and Qb the distribution cost. (Haggett, 1968. p. 144). When V is greater than 1.00, the orientation of industry would be to the source and when less than 1.00, it is to the destination. In n-point situation which is more common, various pulls are at work and one with the highest transport cost generally succeeds in drawing the industry to itself, so that its distance is minimum.

Last few years witnessed an awareness in industrial field and that is about industrial efficiency, which lies in elimination of all wastes. Efficient location of industry forms part of it and it partly means elimination of unnecessary cost of transport by minimisation of distance. Spatial efficiency in location of industry however lies, along- with minimum distance, in maximisation of resources as well as market potential.

Industrial Location Theories

The central aim of industrial location theories is to find an optimal location for industry that gives highest profit with minimum cost. They assume man to be an optimiser. In fact, he often exhibits satisfying behaviour rather than optimisation. So also, it is not possible for an optimum location to remain so, because the conditions change. Many carefully selected locations, therefore, would turn out to be sub optimal and yet the firms run successfully.

In a free economy, there are two approaches to the consideration of location of industry (1) least cost approach and (2) maximum revenue approach. Accordingly, the location theories can be divided into two major groups.

The Least-Cost Theories

These theories imply perfect competition but hold market demand constant. The only means of profit maximization is by way of reducing the transportation costs of both inputs and output. The problem of finding such a least-cost location for an industry was treated from the 19th century with models which were generally simple construction and their objective function was minimum transport. Launhardt developed a location triangle with two sources and one market. Schaffle suggested a gravity model for industries to locate in or near larger towns. In the 20th century, Weber (1909) was deeply impressed by Launhardt's work and produced his renowned theory of location of industry based on minimum movement principle. As Alexander (1977, p. 620) put it, the person to work up a comprehensive theory accounting for location of manufacturing was Alfred Weber.

Alfred Weber

Weber's theory seeks out that location for industry which minimizes movement costs for procuring raw materials and distributing the finished products. He suggested that the transport costs were determined by the weight of the materials and product as well as by the distance to be covered by them. The combination of weight and distance produced his simple index of cost, the ton/mile or km. The locational problem is, therefore, to find the point where the total ton-mileage is minimized in the production distribution process. He used this index, called 'material index', to divide industry into material or market oriented according to weight-losing and non-weight-losing or weight gaining materials. Weber is well-known for his use of isodapancs of equal transport costs, based on isotims of assembly and distribution costs (Fig. 5.1), to find graphically a point of minimum transport cost as an optimum location for industry in n-point situation. But he recognized two circumstances under which location could be attracted away from the least-cost point namely, cheaper labour and agglomeration, as the savings therein exceeded the extra transport costs involved. Weber's model, then, identified fourteen hypothetical types of industry with their different orientations (Hamilton, 1969, p. 371).

Weber was much criticized for an unrealistic framework and his emphasis on the least-cost location as the maximum profit location. Weber failed to identify the revenue aspect of the industry's operation, while concentrating on the minimum costs (K. Siddartha, 2000, p. 157). However, the criticism did not reject his thesis outright, but it paved the way for a shift in thinking of industries, having a revenue motive also, so as to tap the largest sale potential in the market area.

Maximum Revenue or Market Area Theories

These theories assume market as a large area rather than a point in which industries are located and compete to control as much area as possible. McCarty and Lindberg (1966, p. 197) call it competitive marketing which refers to the selection of locations with a view to maximise revenues rather than minimizing costs. The consumers and their behaviour get more importance in the locational decisions. Watts (1968, p. 294) has also asserted that the human factor is becoming more important in the location of industry. Weber considered only distance to one market and not its demand, but there are many markets with varying demands, exerting their own pull on the location of the industry.

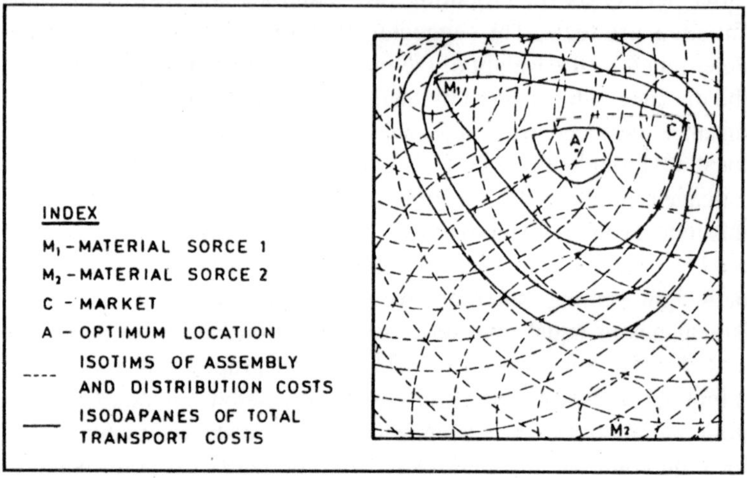

Fig. 5.1: Minimum-Movement point for Location of Industry (Weberian Model)

The spatial economists like Palander, Hoover, Losch, Smith etc. who put forward their ideas on the location of industry, considered both demand and distance and tried to delimit the market areas. In 1940, August Losch demarcated hexagonal market areas of breweries in Germany, with a brewery in the centre of each. It was an efficient spatial arrangement for both the producer and the consumers just like Christaller's in respect of the Central Place theory. David Smith illustrated the maximum profit location for an industry as a composite effect of costs and revenues (K. Siddartha, 2000, p. 172). According to Alonso (1964, p. 338), the best location of a market-oriented industry which minimizes the sum of the distances to its customers is median, which is the point of minimum access to the greatest number of customers.

Now-a-days, input-output analysis is becoming popular among geographers like location-allocation models. Beckmann and Marschak (1955) used the assignment model to allocate optimum location to a plant among many markets and potential suppliers. Recent models are becoming more complicated so as to be applicable to varying situations, more stochastic to achieve greater realism, and more of a satisficer kind than optimal. There is a trend to consider consumer behaviour rather than rigid economic optimacy. Man in his decision-making is not always a rational, economic being and is likely to choose only a satisfactory location to the best of his information and ability. Allen Pred (1967) devised a behavioural matrix for analysis of locational decisions in operational environment. There is also a welfare approach of Reisser and others concerning the prosperity of society as a whole dismissing profit maximization as a motive.

With rising importance of market as the determinant of location of industry alongwith cost of transportation, actually spatial efficiency is being sought even if it is not so called.

Spatial Efficiency in Location of Industry

The trend in model-building of industrial location from minimum movement to maximum revenue has brought into focus the demand or the consumer factor. It is very much akin to the principle of spatial efficiency which relates minimization of distance or transportation cost to maximization of resource and market potential. Both the determinants are required for efficient location of industry and neither is the sole determinant of it unless the other happens to be reasonably constant. Losch (1954, pp. 28-29) also asserts that seeking the place of least cost is as absurd as to consider the point of largest sale as the proper location. Greenhut (1956) has introduced the concept of minimax location combining minimum cost with maximum revenue. His minimax model using game theory, includes different costs and revenues at alternative locations to choose a probable or satisficer location.

An industry is a profit-making business and profits could be earned by minimizing transport costs and maximizing revenue. That is why an industrialist has to carefully choose a spatially efficient location for the industry. The manufacturing sites the world over bear evidence of efficiency in their choice. The great industrial belts of the Western Europe and north America came into being on the basis of material and power resources at hand and large markets nearby generating demand. In India, the industrial regions of Bombay-Pune, Damodar valley-Kolkata, Bangalore-Coimbatore, Ahmedabad-Vadodara and Delhi developed on the basis of large urban markets in the vicinity and resources in the hinterland. The Tata Iron and Steel Company (TISCo) at Jamshedpur, the very first of heavy industry in India, presents a good example of consideration of efficient location. There was another competing site near Chandrapur in Maharashtra State for the industry to be set up. Iron and steel industry is basically resource-oriented and high quality iron ore and coal were available at this site also. But in absence of adequate water and mainly market, it lost to Jamshedpur which later proved to be an efficient location.

Continuous technological progress had an indisputable effect on the location of industry in the 20th century. Cheaper and faster transport and advanced mechanization added to the competitive capacity of a producer in the market, and increased the flexibility of location choice with market becoming more influential factor. Even the material-oriented industry like the iron and steel is having its rolling mills split from the main plant and located near the market. The pulp plant also got separated from the paper mill in the developed countries with different locations—the former remaining at source and the latter moving to the market. These changes in the modern world have to be taken into account while applying the themes of localization to the industries.

However, underlying all trends and changes, the economic motive of profit-making or in spatial sense, that of spatial efficiency remains unchanged. Estall and Buchanan (1968, p. 15) also opine that location choices can change greatly, but the central problem of finding the location that gives the best combination of advantages remains the same.

Spatial Efficiency Measure

As usual, there are two maximum-minimum parameters to use in the measurement of spatial efficiency of industrial location. Each source and market has its own "pull" according to its potential which is a maximising factor, combined with distance or transportation cost which is a minimizing factor on the location of industry. It will be drawn towards the higher value. In case of material-oriented industries, the maximum value of the resource potential among the vying sources, and in case of market-oriented industries, the maximum value of market potential among the vying markets is primarily sought, alongwith the distance or ton/mileage to the sources or markets. Therefore, first of all, an orientation index or Weberian material index may be used to distinguish between the material and market orientation of the industry in point. However though Smith (1955) has tested the efficiency of Weber's index with respect to 65 British industries and found the method imperfect (Haggett, 1968, p. 145), it can serve the purpose of gross division. Then, the spatial efficiency measure can be applied to it to find out its efficient location among vying sources in case of material oriented industry and that among vying markets in case of market-oriented industry.

In respect of material-oriented industry, there are various resources or inputs used in varying quantities and having different costs. The costs of each per unit of production multiplied by its assembly cost or transportation cost according to its weight from each source gives its input index or net distance input for each source. The higher value gravitates the industry to itself and that source may be called spatially more efficient location.

In respect of market-oriented industry, it is a multi-market problem to determine which place has more market potential and stronger pull on the industry. Population of each urban market is presumed to be equivalent to its demand when the actual demand is not known, even though population is not always a surrogate of demand (Smith, 1974, p. 134). The objective function is to find the maximum market potential among the competing large urban centres as markets with the formula:

$$MPi = \frac{Pi}{\sum_{\substack{i=1 \\ r=1}}^{n} w\,dir}$$

Where,

MPi = Market Potential of place i, Pi = Population of place i, and wdir = weighted distance from i to source r (assembly of resources or net distance input).

Spatial efficiency can be determined for each urban centre according to its own pull of market potential and the pulls of market potential of other vying urban centres. The formula is:

$$SEli = \frac{MPi}{\sum_{\substack{i=1 \\ j=1}}^{n} MPj/dij}$$

Where,

SEIi = Spatial efficiency of industry at place i,
MPi = Market potential at place i, and
dij = Distance from i to j

The Ballarpur Paper Mills

The spatial efficiency measure has been applied to the location of the Ballarpur Paper Mills at Ballarpur in Chandrapur district of Vidarbha (Fig. 5.2). The general trend in the paper industry is to separate the pulp and paper plant. But the Ballarpur Paper Mills is an integrated pulp and paper plant and it is ought to be material-oriented as the vegetal matter used as raw material is bulky and three-times weight-losing. Therefore the proximity of raw materials is essential.

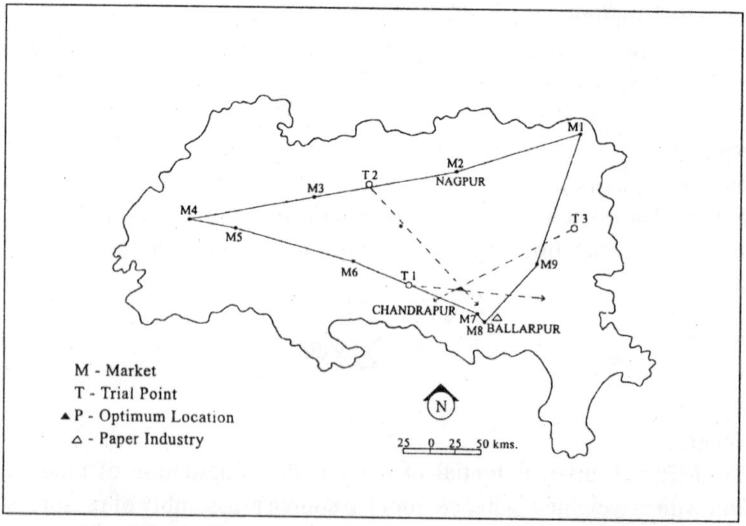

Fig. 5.2: Optimum Location of Paper Industry in Vidarbha

The Ballarpur Paper Mills is renowned in paper production. The raw materials used for pulp are bamboo, wood, grass especially sabai grass from the forest, and soyabean and cotton seed oil, rags and yarn waste in small quantities as farm produce. The present location of the mill is at the source *i.e.* at the margin of vast and dense forest of Chandrapur-Gadchiroli districts. Other inputs like labour are locally available. The chemicals are produced by the chemical division of the same company at the same place. Coal is near at hand from the Ballarpur coal mines. Thus, Ballarpur is in excellent position for assembly of inputs at one place. There are 3 other forest areas in Vidarbha-in Bhandara district (area 156402 ha), in Nagpur district (60542 ha), and in Amravati district (55021 ha) as vying sources for paper industry. But their forests are not only smaller in area as compared to that of Chandrapur-Gadchiroli districts (area 476533 ha), but also inferior to it. So also, Amravati and Bhandara districts do not have any coal. The source location of this material-oriented industry at Ballarpur may be taken as spatially efficient and the industry there is running successfully. The total turnover of the company in 2003 was 2150 crores of rupees as against that of 1615 crores of rupees in 2002.

However, there is another aspect to be considered in the problem, *i.e.* separation of the pulp and paper production. Supposing that the material-oriented pulp plant is efficiently located at Ballarpur but the paper mill which is market oriented has to be shifted, the problem is to find a spatially efficient location for it among competing market places. Some 10 urban centres including Ballarpur, itself a city, are selected as markets from Vidarbha and their population is considered for computation of market potential. The resource for paper production is pulp. So, the distance from each market to pulp producing plant at Ballarpur is used in the computation of market potential. The combined effect of the market potential of each market place and its distance to all others determines the pull of each market place and its spatial efficiency. Higher the value, more efficient is the place for location of paper industry.

The Table 5.1 reveals that Ballarpur is the most efficient location for the paper plant also, since it has its own market and pulp is not required to be moved. Its adjacent city of Chandrapur ranks next and the largest city of Nagpur in Vidarbha comes third in spatial efficiency. Thus, the pulp and paper plant at Ballarpur is assessed to be spatially efficient in its location and it is prosperous there. If two more paper plants are required to be set up, Chandrapur and Nagpur are highly competitive and prospective places for location of the industry, though they are sub optimal. Spatial efficiency has been derived with the help of formulae given earlier.

Table 5.1: Spatial Efficiency of Markets for Location of Paper Industry in Vidarbha

Place	*Population*	*Market Potential*	*Pull of other markets*	*Spatial Efficiency*
1. Akola	3,28,034	1286.4	17416.5	0.0738
2. Amravati	4,21,576	2129.2	21111.7	0.1008
3. Ballarpur	83,511	8351.1	38517.7	2.1681
4. Chandrapur	2,26,105	2055.5	17835.6	1.1524
5. Gadchiroli	29,565	394.2	23082.0	0.0171
6. Gondia	1,09,470	561.4	19268.2	0.0291
7. Khamgaon	73,692	245.6	19575.1	0.0125
8. Nagpur	16,24,752	11441.9	10521.6	1.0875
9. Wardha	1,02,985	858.2	14185.2	0.0605
10. Yavatmal	1,08,578	781.1	27501.9	0.0284

The optimacy of the Ballarpur paper plant is further tested graphically with the help of vector traverse in Fig. 5.2. (Haggett, 1968, p. 147). Three trial points are chosen with respect to the bounding polygon of market places from which vectors in proportion to their spatial efficiencies are drawn and cumulated to form three composite vector traverses. The end of each vector-traverse is joined to the respective trial-point by broken lines. These lines cut each other to form a small triangle where all the pulls are balanced and the actual most efficient point lies. As the Fig. 5.2 shows, it is a little northward (may be due to greater pull of Nagpur) but still near Chandrapur-Ballarpur. The existing paper mill at Ballarpur, therefore, is sub optimal location, but has more prospects than any other market place to be a successful and satisfactory location.

Conclusion

In this fast industrializing world, it is essential to find an efficient location for the industry so that it can be viable and make profits.

The problem is spatial-economic and not a simple one because there are so many factors involved. However, spatial efficiency of industrial location boils down upon demand and distance. Earlier models like Weber's emphasised the distance factor but later on, revenue and market area were given due importance. Spatial efficiency model also considers market potential as well as distance or transportation cost, in primal-dual location problem. More intricate quantitative methods like linear programs or game theory may be employed to obtain more refined results of optimacy which would be reliable if both demand and distance are duly considered.

Some industries, say textiles, may be located at more than one place in a region, even at several places in a belt and all of them cannot be optimum. It is common that sub optimal locations are chosen for industry on satisficer level for one reason or another and yet they do good. Haggett (1968, pp. 151-52) cites an example on Cotterill's (1950) study of gas belt of Texas and Illinois smelting industry, out of which the former is in much more advantageous position. But the latter survives on the sale of by products and turns deficit into profit.

A unique location of industries at port cities everywhere in the world combines the advantage of export and import, and many ports like Nagasaki, Nagoya, Bombay, London, Boston, Detroit, Chicago etc. have become industrial giants. Empirical studies of such cases are required to bring out various locational advantages that influence efficiency.

An interesting geographical feature is the shape of areas which has spatial dimension and its own peculiar efficiency. The next chapter deals with that.

REFERENCES

Alexander J.W. (1977), *Economic Geography,* Prentice-Hall of India Pvt. Ltd. New Delhi, p. 620.

Alonso. W. (1964), Location Theory, in L. Needleman (ed), *Regional Analysis,* Penguin Books Ltd. Harmonds Worth, England, p. 338.

Brisco N. (1914), *Economics of Efficiency,* The McMillan Company, New York, p. 53.

Chisholm M. (1968), *Geography and Economics*, G. Bells and Sons Ltd. London, p. 2.

Estall. R.C. and R.O. Buchanan (1968), *Industrial Activity and Economic Geography,* Hutchinson University Library, London, p. 15.

Ibid, p. 15.

Haggett. P. (1968): *Locational Analysis in Human Geography,* Edward Arnold (Publishers) Ltd., London, p. 145.

Ibid, p. 147.

Ibid, pp. 151-52.

Hamilton F.E.I. (1969), Models of Industrial Location, in R.Y. Chorley and P. Haggett (eds), *Socio-Economics Models in Geography,* Methuen and Co, London, p. 371.

Losch A. (1954), *The Economics of Location,* Yale University Press, New Haven, pp. 28-29.

Mc Carty H.H. and J.B. Lindberg (1966), *A Preface to Economic Geography,* Prentice-Hall, Inc. Englewood Cliffs, New Jerrey, p. 182.

Ibid, p. 197.

Siddartha K. (2000), *Economic Geography: Theories, Processes and Patterns,* Kisalaya Publication Pvt. Ltd, Patna, p. 157.

Ibid., p. 172.

Smith R.H.T. (1974), Concepts and Methods in Commodity Flow Analysis, in M.E.E. Hurst (ed) *Transportation Geography,* McGraw Hill Book Company, New York, p. 134.

Watts D.G. (1968), Changes in Location of the South-Wales Iron and Steel Industry 1860-1930, *Geography*, Vol. 53, Part 3, p. 294.

6

SPATIAL EFFICIENCY IN SHAPES AND BOUNDARIES

Boundaries are political as well as administrative dividing lines, delimiting the areas of authority and organisatoin and at the same time separating territories and people. Both physical and human elements are involved in boundary formation which may become, at times, sensitive.

Boundaries give rise to a peculiar geographical feature—the shape of the region. An area has shape when its boundary is a closed curve that circumscribes a two dimensional surface (Massam, 1975, p. 14). To understand the spatial organisation in an area, its shape or form need be analysed. However, shape studies of areas are not much advanced. Bunge, Blair and Bliss, Boyce and Clark, Haggett and Chorley, Boots, Gibbs and Whittington have produced some work on shape measures.

Shape has a spatial existence in the form of a territorial unit which is created by boundaries located in space. Their location is closely related to spatial efficiency. Spatially efficient boundaries are those which are short and clear-cut so that there arise minimum boundary problems. To keep the boundary uncontroversial either it is generally drawn, as far as possible, along an existing physical divide such as a ridge or a river or a cultural divide such as roads, railways or language/religion differences or kept straight. This is true of internal as well as international boundaries of a state. They are frequently made to coincide with linear physical or cultural features (Prescott, 1965, p. 69). As the boundary has its efficiency, the shape of the bounded area also exhibits spatial efficiency.

A peculiar thing that stands out in case of boundaries is that they are seldom straight unless planned that way. On the other hand, they are very irregularly curving, giving a weird shape to the enclosed area. It becomes difficult, therefore, to measure such shape patterns and define their spatial efficiency.

Measurement of Shape

Shape is difficult to measure and there has not been much work on the subject. Geographical pattern and shape are different things though pattern may have shape characteristics. Both are not related to the size of the area.

One of the interesting features of shape and boundaries is called the contact number (CN) which is defined as the number of neighbours, touching the boundaries. If the mean CN for sets of areas is calculated it comes to about 6, suggesting Christaller's hexagonal shape in packed distribution. Meyer and Huggett (1979, p. 38) remark that the ideal shape for packed territories in theory is hexagon.

Shape is essentially measured for its compactness and most compact shape is represented by a perfect circle, because the circumference is smallest as compared to the area it encompasses (Fig. 6.1). The circle has the least circumference and the radial distance. In the lattice, the hexagon ranks next to it.

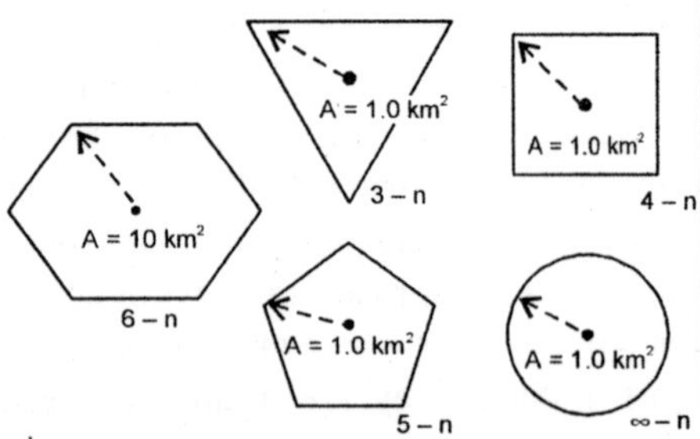

Shapes	Area Sq.km	Circum km.	Radial dist.km.
Circle x-n	1	3.6	0.20
Hexagon 6-n	1	3.8	0.62
Pentagon 5-n	1	3.9	0.65
Square 4-n	1	4.0	0.71
Triangle 3-n	1	4.6	0.88

Fig. 6.1: Efficiency in Geometrical Shapes

So the circularity and the compactness of a unit area is measured by transforming the given area into a circular shape and then comparing its circumference with the real boundary length. More the difference between the two, greater is the deviation from circularity and compactness of the shape. Thus, France has a value of two, indicating almost ideal shape, while Norway has a value of eleven, indicating elongation (Valkenburg and Stotz, 1957, pp. 59-60) J.P. Gibbs (1966, pp. 99-106) has given a method of shape measurement which is also comparable to a circle. He recognises four basic shapes *i.e.* circular, rectangular, star and elongated, which could be ascertained precisely by a simple measure. Using a few basic parameters such as area, perimeter, length of the longest axis etc. he has devised shape indices. A circle has shape index of 1.00 or 100 per cent and the indices of other shapes drop down progressively as they deviate from the circle. Meyer and Huggett (1979, p. 118) agree that a circle yields an index of 1.00, a hexagon 0.75 and a long narrow strip 0.1 (Fig. 6.1) Higher the value, more the shape is near a circle. The formula for shape measurement as given by Gibbs (1979, p. 101) is:

$$MC = \frac{100Aa}{(3.1416)(Dp/2)^2}$$

Where,

Mc = Measure of circularity, Aa = Actual area of the unit and

Dp- = Distance between two farthest points on the periphery.

Bunge has also suggested a method of building an equal-sided polygon on the area to measure its shape. The method is complex but it does measure certain extreme shapes with greater accuracy and it does not require the prior determination of a center (Yeates, 1974, p. 39). The Boyce-Clark method is less precise but simpler than the Bunge's. It is based on a formula (Yeates, 1974, p. 336)

$$SBC = \sum_{i=1}^{n} \left| \frac{ri}{\sum_{i=1}^{n} ri} 100 - \frac{100}{n} \right|$$

SBC = Boyce-Clark shape index, and ri = length of the ith radial.

The major problem is to determine the central point from which the radials are drawn. Boyce and Clark suggest the centre of gravity of the area, but it can be any meaningful centre. The radials are equally spaced from this centre to the perimeter and their length is measured. If SBC is zero, the shape is a perfect circle. The index for a hexagon is 7.7 for a square near about 12, for a cruciform 18, for a star shape 25 and for a rectangle 28. As the elongation increases, the index also increases, for example, to 175 for a strip. Thus, various shapes are quantified and their interpretation becomes easy.

Spatial Efficiency of Shapes and Boundaries

The boundary and the shape of the demarcated area go together and measurement of spatial efficiency involves both of them. Ideally a boundary line should be regular as far as possible. But in reality most of the boundaries are quite indenting and irregular. Wherever they could be planned, they have been kept straight such as the internal state boundaries in Australia or the boundaries between the USA and Canada except in the Great lakes-St. Lawrence region.

An efficient shape is one which has maximum area as compared to its perimeter or boundary length and the minimum radial distance. It gives the area compactness. The shape efficiency is thus, based on its compactness which is favourable for internal organisation and integration as there is maximum possible closeness of the places in the area. According to Van-Valkenburg and Stotz (1957, p. 59), compact shape is an advantage for internal unity and organisation of the state.

In regular polygons a circle which is the most compact shape is spatially most efficient and so is the circular boundary. Since it is the most ideal shape, the measurement of efficiency of any other form, regular or irregular can be done in terms of its deviation from circularity.

The second best shape is a hexagon in a lattice. For contiguous areas, where circles do not fit with each other, as in case of service areas of the central places, the hexagonal shape and the straight-line boundaries should be taken as efficient.

A shape index which measures circularity also indicates efficiency of shape. It has been computed for India and Nigeria (Fig. 6.2) with the help of the Gibb's formula (1966, p. 101) $MC = \dfrac{100Aa}{(3.1416)(Dp/2)^2}$.

The results are tabulated below:

Table 6.1: Shape Index

Country	*Shape Index (MC)%*
India	29.40
Nigeria	60.47

Both India and Nigeria have quite different values, deviating differently from circularity, *i.e.* 100 per cent. India has a star shape, far away from circularity but Nigeria is closer to it. Therefore, Nigeria is spatially more efficient than India as far as shape is concerned.

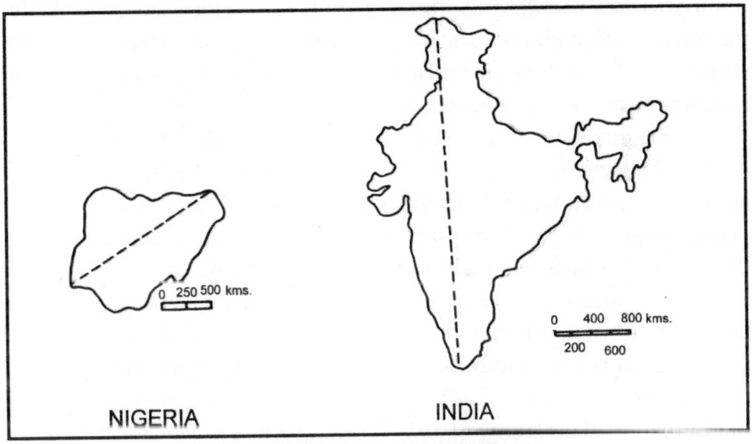

Fig. 6.2: Shapes of India Nigeria

The shape index of transportation network (π) is derived by the formula $\pi = C/d$ where C is the total length of the network and d is the diameter *i.e.* the longest distance by the shortest path (Haggett, 1968, p. 240). It also measures circularity and therefore spatial efficiency. The developed countries like France with a good network of routes have a fairly high value while the developing countries with poor route connections like Bolivia lag far behind with low value.

Is there a correlation between the development and an efficiency in shape of a territory? It is there as far as the development is concerned with the internal organisation of the state. For, an efficient shape is good for internal governance and organisation and perhaps lesser boundary disputes. It may thus help in development process. However, with the revolution in the filed of information and technology, making way for unprecedented communication facility, this advantage will go on diminishing in the times to come.

Location of the Capital or the Headquarters

The shape of a region involves the location of its capital city or the administrative headquarters. Ideally, its location should be in the centre of the region, which is the least distance point, most accessible for the whole area and therefore the most useful place for internal organization and governance.

A central location is advisable for a state capital from defense point of view also. Many capitals generally have a location quite inside the territory, though not exactly in the centre. Capitals have also been shifted inside, to Ankara in Turkey, to Ottawa in Canada. A central location may not really be efficient for the capital if the large part of the state is inhospitable in some way or the other. Moscow is not in the centre of Russia which has a vast poorly populated Siberian land. In such cases, population gravity centre is more acceptable than a geometrical centre for the capital which is efficiently located if it is in the midst of population it administers. Some capitals like London have a peripheral location, not even inside the territory. But there are historical and cultural reasons behind it. The same is true of Delhi, the capital of India, which is also not in its centre, though quite inside the territory.

Jagtap (1983, pp. 377-78) has used population gravity centres for measurement of centrality of district headquarters in Maharashtra state. Krishnan (Surya Kant, 1988, pp. 23-24) suggests that in Indian setting, headquarters of a district must be within one hour of bus journey from the most distant point. Population gravity centre is found by the centrographic method (chapter 3) which yields the mean centre of point distribution of population with the help of x and y co-ordinates. A capital or a headquarters is spatially efficient if it is located near the population gravity centre of the region because it is then at a minimum distance from maximum number of people it governs.

Conclusion

Boundaries are man-made excepting the sea-coast. They have political, economic and ethnic significance and it may trigger boundary disputes off and on. Though the reasons behind such disputes are not mere geographical, short and straight boundaries may help in resolving the issues and keeping peace. Generally, the boundaries are very irregular, following the physical or cultural divide but it is desirable that they should be as efficient as possible.

Boundaries create shapes of areas which are also not regular polygons. Circle is the most efficient shape and indices have been devised to measure the deviation of actual shape from circularity. A shape closer to a circle and efficiency is an advantage for integration and internal organization and hence for governance, though fast developing communication technology is poised to reduce it.

The modern world is witnessing tremendous changes, one of which is the stupendous growth of urban phenomenon. The urban areas have devèloped a characteristic land use pattern which has become a significant feature of landscape. Spatial efficiency in urban land use is the subject matter of the next chapter.

REFERENCES

Gibbs. J.P. (1966), A Method for Comparing the Spatial Shapes of Urban Units, in J.P. Gibbs (ed), *Urban Research Methods*, Affiliated East-West Press Pvt. Ltd. New Delhi, pp. 99-106.

Ibid, p. 101.

Haggett P. (1968), *Locational Analysis in Human Geography*, Edward Arnold, London, p. 240.

Jagtap G.P. (1983), Measurement of Centrality of District Headquarters in Maharashtra State, *The Deccan Geographer*, Vol. 21, No.2 Secunderabad, pp. 377-78.

Massam B. (1975), *Location and Space in Social Administration*, Edward Arnold, London, p. 14.

Meyer I.R. and R.J. Huggett (1979), *Settlements: Geography-Theory in Practice*, Book I, Harper and Row Publishers, New York, p. 38.

Ibid, p. 118.

Prescott, J.R.V. (1965), *The Geography of Frontiers and Boundaries,* Hutchinson & Co. (Publishers) Ltd., London, p. 69.

Surya Kant (1988), A Geographic Perspective on Location of Headquarters of Administrative Areas in India, *Annals of the National Association of Geographers*, Vol. 8, No.2, New Delhi, pp. 23-24.

Valkenburg S.V. and C.L Stotz (1957), *Elements of Political Geography*, Prentice-Hall Inc. Englewood Cliffs, N.J. pp. 59-60.

Ibid, p. 59.

Yeates M. (1974), *An Introduction to Quantitative Analysis in Human Geography,* McGraw Hill Book Company, New York, p. 39.

Ibid, p. 36.

7

SPATIAL EFFICIENCY IN URBAN LAND-USE

Modern times are witnessing a tremendous growth of urbanization and urbanity. Cities are growing so fast and vast that it has become hard for the urban studies to keep pace with the urban phenomenon and a host of problems it is generating. Urban studies and empirical analyses are offering systematic themes and models to interpret the complex nature of urbanity. B.J. Garner (1969, p. 303) remarks that new frameworks are needed if a deeper understanding of spatial organization of this particular aspect of human activity is to be obtained. The concept of spatial efficiency is one of such frameworks to enhance understanding of the urban processes and problems.

Urban studies deal with urban entities from the point of view of their location and function on the one hand and their morphology or internal structure on the other. Or, in other words, it is a study of town in area and town as area. As Carter (1972, p. 8) says, it emphasizes the two main themes of study—the town as a distributed feature and town as a feature with internal structure, or in other words the town in area and the town as area. In the first approach, the towns are considered as discrete phenomena in the general fabric of settlement distribution in an area. As such, they can be analysed geographically in much the same way as any other phenomena similarly distributed. (Mayer and Kohn, 1967, p. 5). The towns are, thus, integrated in the locational structure of settlements as a distributed feature. But at the same time, they also have a wide area of occupance internally, giving rise to an internal structure or the townscape or cityscape as it is called. It is intimately concerned with the personality of the town and it forms another important aspect of urban study.

The town or city carries out various functions, and the urban land is utilized for these functions. Besides, the large urban population, engaged in various urban activities, use the land for its residence which is also taken to be a kind of urban function. These functions, alongwith the streets and routes and open spaces also, give rise to specific land occupance which is known as the urban land use pattern.

Urban Land Use

The urban land is put to various uses to perform urban activities and keep the urban life throbbing. The urban land is given over to commercial, residential, industrial, socio-cultural, professional and administrative functions which interact and vie with each other for the urban land and in the process get spatially organized in a typical fashion through the intra-urban transport system and daily rhythm of movement to produce a characteristic spatial pattern of urban land uses. The urban studies of the internal structure are concerned with the concepts and generalizations about the character and intensity of land uses within the city and with the spatial interactions among them. Harvey (1975, p. 290) also remarks that the complex pattern of land occupance for various uses is focused upon in urban studies to realise processes of spatial organization of urban functions and spatial behaviours of urbanites.

The urban land use patterns are again controlled by man's basic tendency for spatial efficiency. Man strives for maximization of landuse and minimization of distance to the centre of the city which is discernible in the competition for the urban land, land values, and centripetal and centrifugal forces which are so very distinct in the western cities or those of the developed countries, but also prevalent in the cities of the east or the developing world. The resulting patterns denote the level of spatial efficiency reached with the constraints of the given conditions and therefore the principle of spatial efficiency is applicable to the internal organization of land uses in urban areas and their patterns. Now-a-days, planning has become necessary in towns and cities for one reason that there is an unprecedented growth of them and for another that planning brings order and better living, and planning necessarily seeks efficiency for the purpose. To understand the underlying principle of spatial efficiency in the urban land use patterns is important from the point of view of urban planning also.

Most of the urban land that is used for various purposes is built-up area, called as brick and mortar that is used for commercial, residential, industrial, administrative and the like activities. But it also includes open spaces as parks and gardens, playground and even agricultural land known as green belt. The urban streets and roads which are the arteries of urban life, the vital connections between urban activities, form a special feature of urban land use. Traffic on the roads is also a part of urban life and land use interactions.

Urban land use patterns are various but generally speaking, they are found to be closely related to economic development of the societies. They are more or less similar in the developed countries, popularly called as the Western world and those in the developing countries are quite different from their western counterpart.

The Urban Land Use in the West

The developed countries in the west are most urbanized and have developed a land use pattern that follows a definite course. Urban activities occupy different parts of urban land, dictated by the centripetal and centrifugal forces or the forces of attraction and repulsion working at one and the same time. (Colby, 1967 p. 287) They are basic to the processes of centralization and decentralization, concentration and de-concentration, segregation, invasion and succession. These processes cause the urban activities to get spatially differentiated among themselves. The urban activities, first, compete for central location in the city, but the commercial activities get the better of others in the intense competition and occupy the prime location of the centre. This commercial part is called the Central Business District or CBD as a short form. The land is so intensely used for commercial purposes in the CBD that its value soars high and the buildings also rise high. Other activities move outward, especially the high class residences and industries, which prefer certain spatial advantages of large open land and peaceful environment outside, facilitated by fast means of transport. The middle and low class residences do not afford to move as much away as the high class residences but they follow them by succession. The outward movement is incessantly going on in a growing city with the march of the brick and mortar by invasion and succession, resulting in an urban sprawl. Residential areas are, thus, segregated according to class or income of the residents. (Berry and Horton, 1970, p. 443).

These common features of the urban land use in the west are not so well developed in the developing countries, especially in the orient. They have developed the land use pattern suited to their own conditions. Indian towns and cities may represent well the urban land use of the developing countries.

Indian Urban Scene

The developing countries like India lag behind the developed countries of Europe and N. America in urbanization and urban development as well as in economic development. McGee (1975, p. 224) remarks that in developing areas the sequence of events leading to urbanization is perhaps different from that noted in 19th century Europe and N. America.

There are certain typical features of urban places of the developing countries. In the first place, like any developing country, India still is one of the much less urbanized countries of the world. Its level of urbanization was attained by the USA in 1870. Secondly, the total mobility of the townspeople is quite low, much less than that in the Western countries because of the poorer conditions of the people and also poorer transport facilities. Thirdly, many towns have large populations but they lack truly urban living and look nothing more than enlarged villages. Gerald Breeze (1969, p. 99) observes that the grossly low level of subsistence urbanization may be so close to an agricultural subsistence level of living that the difference may be just one of locale.

Spatial organization of urban land uses in India is characterized by a high mix due to their heterogeneous occupance of space and a well-developed CBD is conspicuous by its absence. Johnson (1980, p. 178) calls it patchiness. The so-called commercial core of the town is considerably mixed up with residences and industries. It is quite common to find residential, industrial, and commercial activities, all taking place in the same lot or in the same block. Often a shopkeeper lives at the back of his shop, a convenience that eliminates the journey to work. Even in large million cities, this sort of indiscriminate land use persists (Mishra, 1978, p.11). In her study of Ahmadabad city,

Anjana Desai (1990, p. 525) also finds that various uses of land are mixed in such a way that the place of work gets merged with the place of living. The census of India, therefore, recognises shop-cum-residence and workshop-cum-residence in its housing statistics. Commonly the central core is commercial and inhabited by rich merchants, whereas in the outer periphery the poor people live. The zone between the core and the periphery is inhabited by a mixed population of rich and poor. Neighbourhood enclaves of specific castes and linguistic or religious minorities, formed in different parts of the city, are also a peculiar feature of Indian cities.

Because the central part of the city cannot maximally develop into a purely commercial centre, as a CBD, the commercial activities get concentrated in separate centres outside the core. Trading and wholesale activity is concentrated in Chandni Chowk, Sadar Bazar, Azadpur, Kashmere Gate and Moti Nagar in Old Delhi (Sethi & Pande, 1987, p. 33). Smita Joshi (1987, pp. 38-39) has shown one retail business of drug stores located in the core, outside the core and even near the periphery of Pune. This type of land use by the commercial activities has prevented further the intense utilization of the core for business by skyward construction.

The high class residences generally occupy the land near the centre of the city, though they have now started moving outward, forming suburbs of posh living (Desai, 1990, p. 526). Apart from separate industrial suburbs, there is a tendency for residential areas to be segregated from other uses of land. The new colonies are markedly well planned and having spacious modern buildings, though they are still mixed with rural dwellings and poor houses. In general the rich compete successfully for location near the centre pushing the middle and low-class people outside, though formation of new residential colonies is showing up in urban land use. The poor either have to make daily journey to work on foot or by slow moving vehicles or they give rise to a big problem of Indian city scape, the slums (Jain, 1978, p. 422) by occupying any vacant land in the midst of busy areas.

Another remarkable feature of large cities of India, a remnant of the colonial rule, is the part known as the civil lines, a colonial township beside the old one, having European urban land use pattern with a small CBD, spacious bungalows and administrative quarters. The British have also left the legacy in the form of cantonment towns, railway towns, hill stations and planned cities. McGee (1967) identified three different patterns of growth of cities in the developing countries—the indigenous city, reflecting oriental urbanization, the colonial city, reflecting western elements and modern city, reflecting planned features. However, there is a mixture of all these three elements in several Indian cities. New Delhi, Chandigarh and Jamshedpur are the masterpieces of post-colonial city planning with functionally differentiated land uses, systematic street design, large open spaces, adequate social facilities and the like. But they are exceptions rather than the rule.

The general Indian urban land-use is quite different from the standard suggested by Peter Hall (Khatu & D'Souza, 1982, p. 179) for a city of one million population in developing countries. The following Table 7.1 compares it with national average and that of Ahmadabad city and Nagpur city as examples.

Table 7.1: Urban Land Use in Developing Countries

Land Use	*Ideal for a million city (%)*	*National average of India (%)*	*Ahmadabad city %*	*Nagpur city %*
Residential	14.8	47	42.8	42.0
Industry	14.7	10	17.3	3.0
Commercial	5.5	3	2.0	2.2
Open space	14.7	4	9.8	1.8
Public & semi public		12	4.2	25.0
Roads, Rlys & Airport	25.0	13	11.3	18.2
Others	25.3	11	12.6	7.8
Total	100.00	100.00	100.00	100.00

Source: Khatu and D'Souza, 1982, p. 178.

Source for Nagpur: Nagpur Municipal Corporation.

The preponderance of residential land use in India suggests congestion. The percentage of open spaces for parks, gardens, playgrounds is also very low, pointing again to congestion. The percentage of land under the commercial use is drastically low together with the fact that there is no upward construction of building more than one or hardly two storeys to intensify the uses. It means that the commercial activity is not so intense as to form a distinct CBD and develop the city on the western lines. Fig. 7.2 shows the distribution of land uses in Nagpur city which reveals the prominent features of Indian urban land use. Since the urban land use pattern in India does not resemble that in the developed countries, the land use models constructed for the latter, are not applicable to the former. Yadav's (1979) application of Hoyt's sector theory to the city of Delhi reveals a gross inapplicability of the model.

The Street Pattern

The streets and railways, form an important part of urban land use which provide the necessary links for intra-urban interaction. As such, they may be aptly called the skeleton of the townscape that evolves along with the town's growth. Broadly there are 3 main types of street pattern: (i) The rectangular or the grid; (ii) the radial concentric; and (iii) the irregular. (Dickinson, 1964, p. 167). Other varieties are but the combination of these three. They are grossly represented as the grid by the American cities, the radial-concentric by the European cities; and the irregular by the developing or oriental cities. There is a recent addition to the pattern by the construction of fly-overs, especially in the developed countries.

The street pattern is crucial for the traffic that flows over it. Now-a-days, the intra-urban traffic on the streets is increasing so fast that the streets one falling short to handle it and various urban traffic problems are cropping up. These problems are closely connected to the nature and pattern of the streets. The nature of the streets, including the width, the surfacing and the conditions of them, along with their pattern, deserves due consideration in the traffic problems. Another newly developed issue which is related to the streets and traffic, especially in the developed countries and raising its head in the big cities of the developing countries as well is the provision of the parking lots. Tolley and Turton (1995, p. 189) note an interesting point in provision of new roads and new parklots with huge investments that they lead to more traffic congestion and by becoming enshrined in the structure of the city, perpetuates the same inefficiency.

The Urban Land-Use Models

Empirical studies of a number of western cities have yielded general urban land use models. As Harris and Ullman (1967, p. 277) say, location types and internal structure are repeated so often that broad and suggestive generalizations are valid, especially if limited to the cities of similar size, function and regional setting.

E.W. Burgess proposed in 1925 that urban land uses displayed a zonal arrangement in concentric circles around the city centre. (Fig. 7.1). Thus, a city develops five zones with the CBD in the centre. Working class is found in the inner zone grading to the highest income group on the periphery. The model is derived from observation of a large number of American cities in general and Chicago in particular (Carter, 1972, p. 163). It is much simplified and idealized, but points out a general trend in the development of land use pattern in American cities. Homer Hoyt did not agree with it and presented in 1939 his sector theory (Fig. 7.1). From a factual examination of rental area maps of 25 American cities, he concluded that apart from the CBD, the rent areas in American cities conformed to a pattern of sectors, growing outward from the CBD in the same direction, rather than that of concentric circles. Thus, the residences of a particular class does not occupy the whole ring around the centre, giving way to another ring, but it occupies the whole sector, radiating from the CBD unto the periphery. The urban land uses are distributed in sectors, developed in different parts of the city, rather than ring like zones. Shafer (1977, p. 133) also agrees that there is a strong evidence for the sector type development of a land use pattern in many cities. Hoyt's model, therefore, may be said to be closer to reality than the ideal Burgess's model.

Still closer to reality is Harris and Ullman's model, developed in 1945 and called as 'multiple nuclei theory'. They argued that during the growth process of cities, separate areas, neither zones nor sectors but nuclei emerged which developed land uses, forming a sort of patchwork (Fig. 7.1). For, towns and cities grow by the progressive integration of a number of separate enclaves which become residential neighbourhoods with small outlying business and other activities within a walking distance of the residents, connected to the core by transport routes. However, they also agreed (1967, p. 285) that most cities combined the aspects of all these three generalizations.

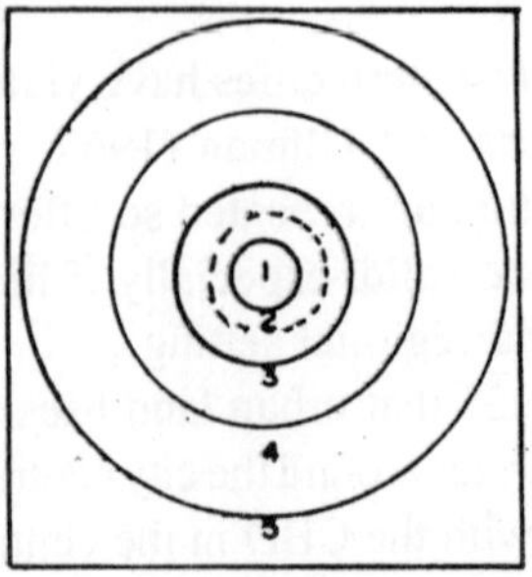

THE CONCENTRIC ZONE MODEL

1. Loop
2. Factory zone
   and
   Zone in transition
3. Zone of workmen's homes
4. Residential zone (single family dwellings)
5. Commuters zone

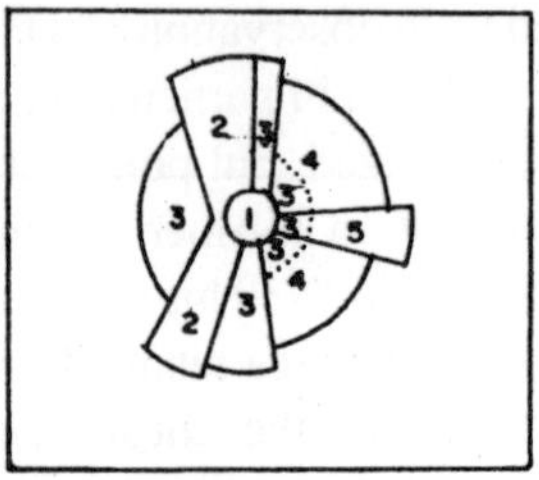

SECTOR SCHEME MODEL

1. CBD
2. Wholesaling & light manufacture
3. Low class Residential
4. Middle class Residential
5. High class Residential

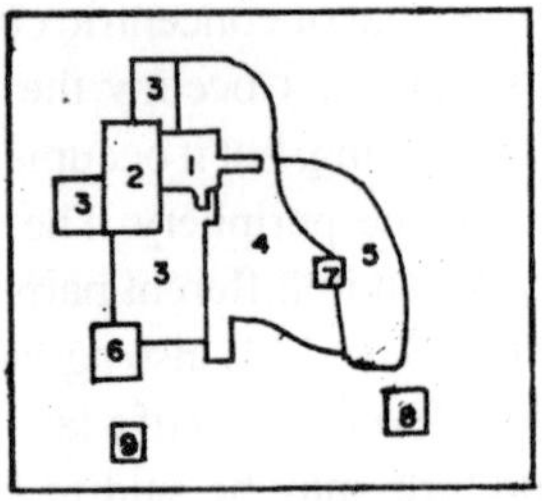

THE MULTIPLE NUCLEI MODEL

1–5 as in sector scheme
6. Heavy manufacturing
7. Outlying business district
8. Residential suburb
9. Industrial suburb

Fig. 7.1: Models of Urban Land Use

These three well-known models of urban land use are developed from and are applicable to the cities in the developed countries in the west. The cities of the developing countries in the orient which intrigued both oriental and occidental scholars, have heterogeneous land use and they do not display the features outlined in these models. Indian cities, representative of the orient, show a patchwork of different activities with outlying business centres in the residential neighbourhoods within the walking distance of the people as Harris and Ullman have allowed for in their model. But a full fledged CBD and residential and industrial suburbs with efficient transport system are absent in majority of Indian cities except a few metropolitan ones.

Spatial Efficiency in Urban Land Use Pattern

Urbanity is distinguished on the basis of secondary, tertiary and even quaternary functions which occupy urban land in different parts of town or city. Urban population is engaged in these urban activities and it occupies much of the urban land for residence. Thus, urban land is used for different purposes and a distributional pattern of urban land uses results from a multitude of decisions made by individuals about location (Garner, 1969, p. 335). These decisions depend upon individual space preferences, competition for space, economics of land values, centripetal and centrifugal forces and intra-urban transport system. Above all, there is a motivation to attain spatial efficiency, working in its subtle way. Therefore, even though each town and city develops its own personality, there are several similarities in the land use pattern according to the level of spatial efficiency reached. Spatial efficiency is reached by the best use of land which yields highest return with minimum possible distance for the people. Shafer (1977, p. 139) comments that there is probably no single concept widely accepted and used in the determination of the uses to which land within urban areas is put than the concept of highest and best use. The cities in the developed and the developing countries have evolved different types of land use pattern in urban areas according to their own socio-economic conditions. They strive for spatial efficiency and best use of land suited to the given situation but reach different levels of it.

The Western Cities

The standard urban land use pattern has been derived from the studies of the cities in the west or the developed countries of Europe and North America. The outstanding feature of the land use in these cities is the commercial core, known as the CBD. The centre is the most accessible part of the city *i.e.* at a minimum distance from all the sides and therefore easily frequented. Naturally, all the activities have a central tendency to locate which gives rise to centripetal forces. As a consequence, competition for the space in the centre increases which is reflected in soaring land values. Only commercial activities come out successful as the highest bidder in the competition and occupy the central part of the city. This process is in keeping with the principle of spatial efficiency. The commercial activities can make the most of the locational advantage of the center by maximising the market potential and minimising distance for the customers. Dickinson (1964, p. 38) also opines that goods and services demanded can most efficiently be carried out at central location within clustered communities. The spatial efficiency is further enhanced by fine segregation among commercial activities. Alike activities group together and more specialisation in goods and services takes place so that they get the advantage of agglomeration and the customers get wider range of them at one location. Murphy and Vance (1966) have observed such a segregation in the core of CBD in American cities.

The central part which is at a minimum distance for the people is most intensively utilised by business out lets. A great number of small size outlets are packed in the small area of the center. If the available land falls short of the demand, the buildings rise higher and higher to accommodate the commercial activities, so much so that the sky scrapers have come into being as a regular feature of the western cities. Then high rise buildings point out the maximisation of the space use and preference of the commercial activities to be in the center and at a minimum distance for the clientele or in other words, spatial efficiency..

An outward movement of other activities than commercial is also spatially efficient in their own way. Wholesale trading is just near the center and also are the light industries. The land far way from the center is ample and cheaper and hence preferred for activities such as industrial plants which require large area for use. The high class residences also prefer the advantage of large, open space and peaceful environment away from the central congestion. Thus they make the most of the space potential and the increased distances, however, are made good by the fast means of transport, especially the automobile. With the advent of automobile only, it has become possible for these activities to move far away from the center. The increased distance is, thus, affordable compromised by the car travel for the advantage of spaciousness, near the periphery. These are the centrifugal forces which initiate a movement outward and also cause a segregation among the residences. The residential areas are segregated according to class or income of the residents, (Berry and Horton, 1970, p. 443) who prefer and occupy different parts of the city according to their affordability. The low class people who do not afford to go a long distance away from the center live in small congested tenements near the center. The middle class people take up residence somewhat longer. However, in the urban growth process, an outward movement of all the classes is going on incessantly by invasion and succession. Thus, the maximisation of space potential and cost of distance ultimately decide the type of use the urban land is put to.

A peculiar feature of the western cities is the small retail business centers, located in the residential neighbourhoods, away from the center of the city. They are also spatially efficient being at a minimum distance from the residential neighbourhoods from which they draw their maximum clientele. The urban activities use the urban land efficiently in the western cities.

However, the western cities have developed, broadly speaking, different types of street patterns. The American cities have preferred gird while the European cities have evolved a radial concentric pattern. The latter is spatially more efficient as the route length is minimum as compared to the area served. The modern cities, however, need efficiency of the roads in the form of their good width, excellent surface conditions, a number of flyovers, spacious parklots and bypasses to handle the traffic smoothly rather than only their pattern. A network of metro railway supplements the urban streets as intra-urban traffic carrier, especially for the commuters. Yet, they are failing to be adequate for the ever increasing traffic inspite of all efficiency they could muster.

Out of the 3 well known models of urban land use, which analyse the general features of city space, the most generalised and ideal model of Burgess's spells out an efficient condition (Fig. 7.1). The arrangement of land uses in concentric zones around the center allows for compact and maximum use of land. The activity in each zone is at the same distance from the center on all sides. The central area is most efficiently used for the CBD. The low class, middle class and high class residential zones are successively away from the center, rightly according to economic conditions of the residents. The poor people live as near the center as possible to save distance and the rich people can afford to go to a long distance towards the periphery. This is an efficient situation.

Homer Hoyt's model accommodates, sector type development of residences as a variant. Anderson and Egeland (1974) saw Burgess's concentric zones and Hoyt's model as complimentary and compatible analyses of the outcome of conflict over scarce land. Johnson (1974) in fact, considered Hoyt's model as a refinement of Burgess's. It takes a note of a situation nearer reality. Harris and Ulman's model also recognises another situation of multiple nuclei which is again nearer reality. The small retail business in the residential neighbourhood is really spatially efficient as it caters for its clientele of the neighbourhood at a walking distance. All these three models are based on the historical development of western cities and on the premises (i) adjustment to the factor of distance, (ii) space use, (iii) tendency of segregation and (iv) space preferences.

The Indian Cities

The developing countries like India have developed their own urban land use as an adjustment to the factor of distance and economic conditions of the towns-people. There is no discernible pattern but a high mix of land uses and no model could apply to it. There is no true development of CBD, since the central part consists of not only commercial activities but the residential and industrial also. People live within short distances from the place of work so that they can walk to it and rightly are the Indian cities called the pedestrian cities (Johnson, 1980, p. 178). Thus, people try to minimise distance but, that way, spatial efficiency is not attained because the center could not maximise the market potential and is limited in its range of business. The commercial activities instead of centralizing to draw maximum clientele, are dispersed in the different parts of the city to be at a short distance for the customers of each side. This is inefficient.

Another inefficient situation in the Indian cities is the high class residential occupance of land near the center and the low class people moving to the periphery. These poor people can neither use the large open land at the periphery maximally nor can they be at a minimum distance from the place of work.

However, as India is advancing in economic development, changes are taking place in the urban land use after the pattern of that in the developed world, at least in large cities. Anjana Desai (1990, p. 576) has observed a new trend of centrifugal force in operation, resulting in an out migration of rich families from the core area to suburbs. Residential areas are getting separated from the original mixed pattern of landuse. It throws light on the fact that spatial efficiency in urban land use depends on the economic development of the society, which facilitates maximisation of landuse.

The street pattern in Indian cities is mostly irregular and hence inefficient. The old streets are narrow and winding and their surface conditions are also not good. The new planned roads are grid or circular, wider, better surfaced and more efficient in handling traffic. In large metropolitan cities, metro railway, flyovers, parklots are now appearing as the traffic is increasing on the roads.

Traffic on the urban roads is intermixed and inefficient. There is an interlocking of vehicles moving at different speeds and blocking each other. On the whole, the traffic is slow moving. More often than not, the alleys and lanes are too narrow to carry the vehicular traffic. Brush (1962, p. 59) has described the state of the streets that they are irregular in pattern, narrow and crooked, often without sidewalks and encroched upon by booths and fronts of shops. The side lanes and alleys are usually much narrower and so crooked so as to be almost impassable for wheeled vehicles. The traffic jams have started occurring on Indian city roads but they are mostly due to the interlocking of vehicles moving at different speeds rather than overcrowding of vehicles.

Spatial Efficiency of Land Use in Nagpur City

Measurement of spatial efficiency in urban land use is essential to see locational maladjustments if any. Spatial efficiency is examined in Nagpur city as an example, which is a typical million city of India. It is the largest city in Vidarbha with a population of 16,24,752 and the developed urban area of 8340 hectares, much more than the ideal 5400 hectares as given by the Peter Hall for the million cities of the developing countries. The distribution of landuses is, however, far from the ideal of Peter Hall (Table 7.2). The residential area takes up the largest share of 42.6 per cent far more than that proposed by Hall *i.e.* 14.8 per cent, but near about the Indian average. Other uses like commercial, industrial, open spaces, roads etc. naturally have a smaller share than the ideal. This situation points out to the overpopulation of the city. The internal landuse of Nagpur is presented in Fig 7.2, which also shows a preponderance of residential area.

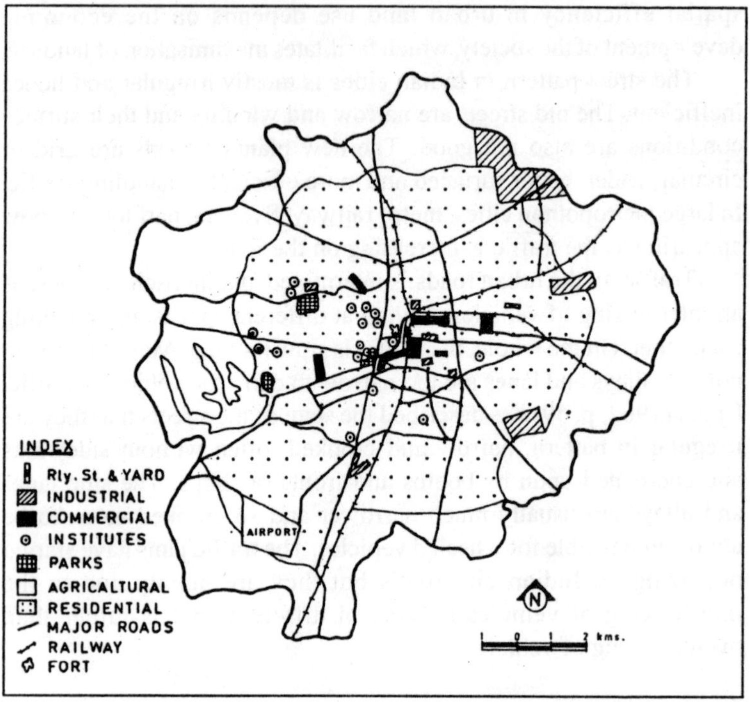

Fig. 7.2: Urban Land-Use in Nagpur City

As Fig. 7.2 brings out further, the land use pattern of Nagpur is quite mixed up, like an average Indian city. There is no development of a hard core of CBD. On the other hand, there are five major market centers including the core, separate from each other, interspersed by residences. These centers are (i) Sitabuldi (the core); (ii) Sadar (to the NW); (iii) Gokulpeth (to the W); (iv) Gandhibagh (to the N E); and (v) Itwari-Mahal (to the E). Besides there are two budding centers at Kamal chowk-Jaripatka and Khamla-Ring Road. All these centers have developed at cross roads and along major streets but they have discouraged the development of a main big shopping center in the core of the city. There are 56,776 commercial outlets but the commercial landuse instead of developing into CBD, is dispersed and look patchy. Apart from these major business centers, small business is also found in the residential neighbourhoods. The dispersion of commercial activity in this way is an inefficient situation since the market potential is divided among different centers and the core area of Sitabuldi is not able to maximise the sale. Rather, it appears that each business center draws limited customers from its own side and be at a minimum distance for them but the commercial activity is intense in none and so, there are nowhere in the business centers, high rise buildings. The wholesale market lies in the eastern part outside the core.

To determine spatial efficiency of five business centers, two parameters, namely space utilisation and distance from the city center, are used. Space utilisation is expressed in land values. Therefore, average land values as shown in Fig. 7.3 and Table 7.3 are considered for computation.

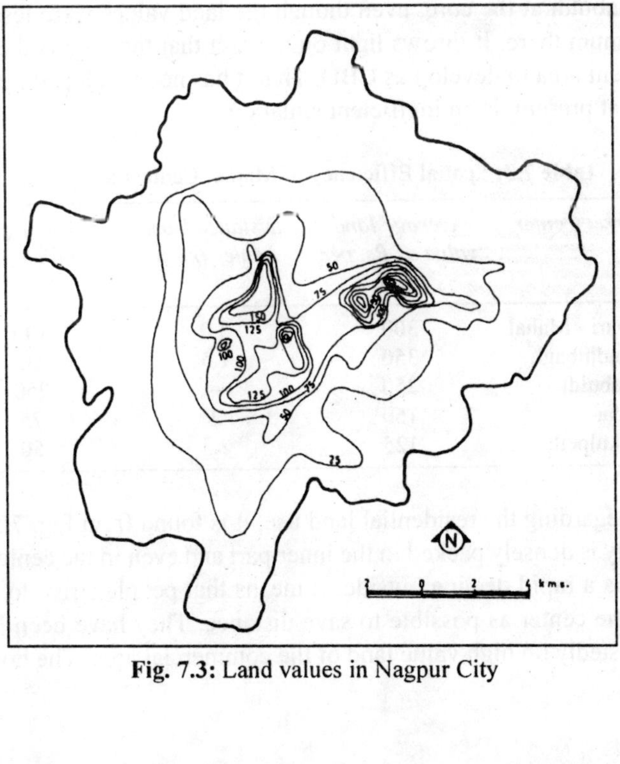

Fig. 7.3: Land values in Nagpur City

As Fig. 7.3 shows, the point of the highest value is not at the center but skewed to the east, under historical influence. Itwari and Gandhibagh in the east are old markets and still continue to be busy shopping areas. They also have an advantage of wholesale trading. The Gokulpeth market in the developing area of the west competes well with the core market of Sitabuldi. To measure spatial efficiency of the market center, the simple formula is

$$SEmi = \frac{LVi}{diC}$$

Where

SEmi = Spatial efficiency of ith market center,
LVi = land value of ith market center; and
diC = distance of ith market from the center of the city.

Higher the value, better is spatial efficiency. The measure is applied to the five markets of Nagpur, the land values of which are of the year 1981.

As Table 7.3 and Fig. 7.4 reveal, spatial efficiency was highest in Sitabuldi at the core, even though the land values were less than maximum there. It throws light on the fact that the core is the most efficient area to develop as CBD. That it has not developed into one even at present, is an inefficient situation.

Table 7.3: Spatial Efficiency of Market Centers in Nagpur

Market Center	*Average land value as Rs. per sq. ft.*	*Distance from center (kms)*	*Spatial Efficiency*
1. Itwari - Mahal	300	3.0	100
2. Gandhibagh	350	1.5	233
3. Sitabuldi	250	–	250
4. Sadar	150	2	75
5. Gokulpeth	125	2.5	50

Regarding the residential land use, it is found from Fig. 7.5 that the city is densely packed in the inner part and even in the center and there is a rapid decline outside. It means that people strive to be as near the center as possible to save distance. They have been living congestedly on high value land of the commercial area. The housing statistics for Nagpur has also identified the categories of shop-cum-dwellings and workshop-cum-dwellings. Because of such urban crowding, there is no development of residential suburbs in a million city of Nagpur and it hampers the maximisation of land use for commercial activity. An inappropriate location of one activity distorts the efficiency of other land uses.

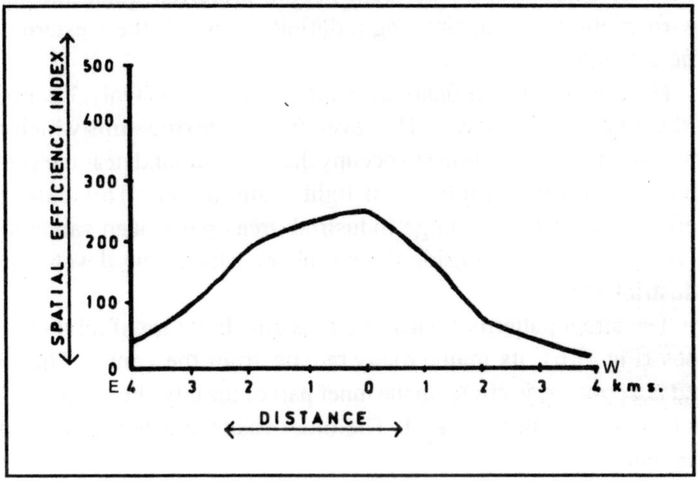

Fig. 7.4: Spatial Efficiency of Markets in Nagpur

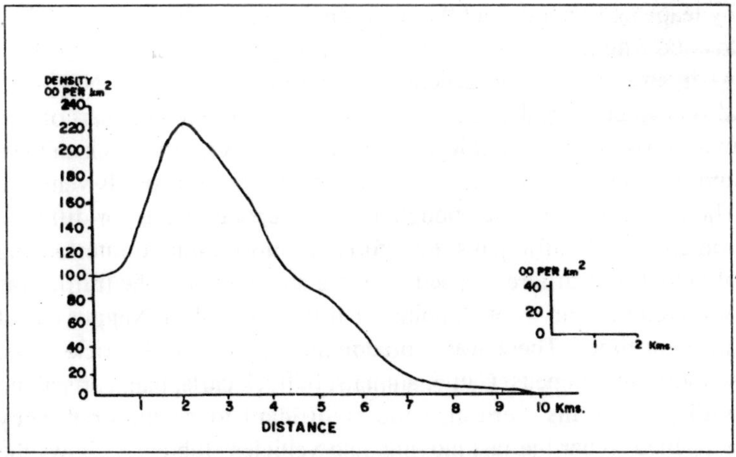

Fig. 7.5: Distance-Density of Nagpur city

The people are now preferring to move outward to new colonies. A survey was carried out of 250 houses in different income groups at regular distances from the center to find out space preference of people for residence. The strong preference of 80 per cent people was to live as near the center as possible until the high rent forbids and they include rich and poor alike, showing no segregation according to economic class. But 20 per cent people preferred quiet and open environment outward, showing a definite trend on the western and efficient line.

Nagpur is not significant in industrial activities. Only 3 per cent of the total land is under it. However, two old textile mills which are now closed down, continue to occupy the valuable land near the center which should preferably be in light manufacture. This calls for inefficiency. However, large industrial areas have been sanctioned by the Govt. at and outside the periphery which may develop into industrial suburbs.

The street pattern of Nagpur city is an admixture of all the three types (Fig. 7.6). Its major roads radiate from the center. The ring road is a concentric circle. In the inner part of the city, the crisscrossing streets have formed the grid. The older roads and feeder lanes are irregular.

Now, in the center, a flyover has been built and some flyovers are also planned. Presently, the traffic on the city roads is increasing by leaps and bounds and the roads and even the flyovers are failing to cope with it, so that a number of traffic problems have cropped up. As in an Indian city in general. The intra-urban traffic in Nagpur is also characteristically slow-moving. It is because a large part of it is made up of pedestrians, bicycles, cycle-rickshaws, carts, pack animals and stray animals, creating considerable nuisance for speedy vehicles. The roads are not wide enough to make separate lanes for different vehicles. The traffic jams, therefore, are mostly due to interlocking of vehicles of different speeds, people and animals. The traffic was observed in a survey at 48 points on different roads in Nagpur city at different times. There was a predominance of bicycles, rickshaws, scooters and mopeds. Cattle, animals, bullock carts, handcarts along with pedestrians were also too significant to be ignored. They happened to bar the fast moving auto vehicles. Table 7.5 shows the average number of auto vehicles stopped by them in an hour.

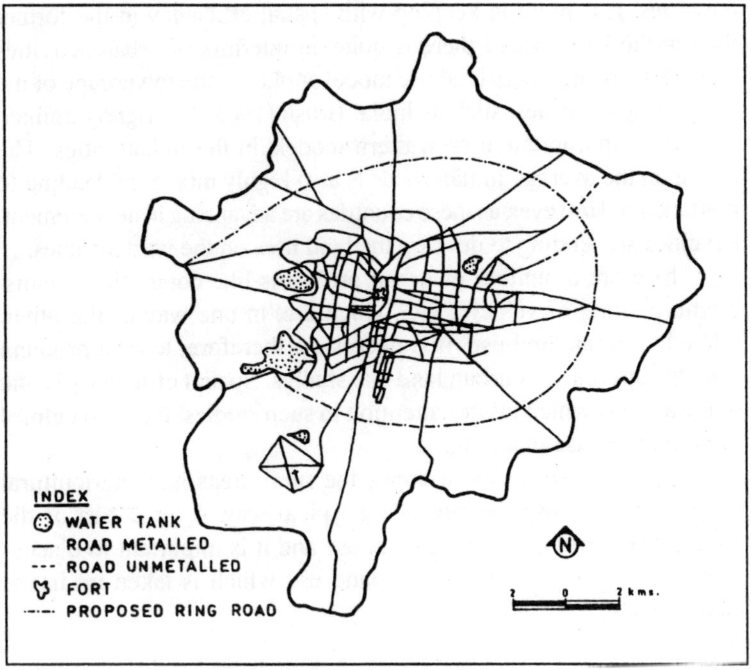

Fig. 7.6: Street Plan of Nagpur City

Table 7.5: Barriers to Auto Vehicles in Nagpur

Barriers	*Average No. of Auto vehicles on a road per hour in Nagpur*					
	Bus	*Car*	*Truck*	*Scooter*	*Moped*	*Autorickshaw*
Padestrians	15	46	15	19	39	17
Cycles	14	27	30	24	29	12
Rickhaws	30	20	22	16	25	13
Carts	23	8	5	10	3	5
Animals	4	10	12	7	7	6

If, this way, the speed of the vehicle drops intermittently, it increases inefficiency of the traffic on the roads which are not maximally utilised by vehicles by covering the distances in shorter time.

Conclusion

Urban land use appears different in the developed and developing countries. It is more in keeping with spatial efficiency in the former than in the later, where there is quite an intermix of urban activities and therefore, no urban land use model applies to the townscape of the developing countries, such as India. Brush (1968) has rightly pointed out the in appropriation of western models in the Indian cities. The traffic on the average Indian roads is also highly mixed up, leading to inefficiency. However, as these countries are advancing in development, the cities are tending to develop the land uses on the western lines.

There are a number of urban problems like congestion, slums, blight, parking of vehicles etc. which are, in one way or the other, related to urban land use. It is necessary, therefore, to take problem oriented approach in urban land use studies. Spatial efficiency is one of the aspects which invites attention in such studies. It is also helpful in urban land use planning.

Like land use in urban areas, the rural areas have agricultural land use which also develops in a typical way. A large part of the land in the world is in agricultural use and it is important to discuss spatial efficiency in agricultural land use which is taken up in the next chapter.

REFERENCES

Berry B.J.L. and F.E. Horton (1970), *Geographic Perspectives on Urban Systems, with Integrated Readings*, Prentice Hall Inc., Englewood Cliffs, New Jersey, p. 443.

Breeze C. (1969), *Urbanisatoin in Newly Developing Countries*, Prentice Hall of India Pvt. Ltd, New Delhi, p. 99.

Brush J.E. (1962), Morphology of Indian Cities, in Roy Turner (ed), *India's Urban Future*, University of California Press, Berkley, p. 59.

Carter H. (1972), *The Study of Urban Geography*, Edward Arnold, London, p. 8.

Ibid, p. 163.

Colby C.C. (1967), Centrifugal and Centripetal Forces in Urban Geography in H.M. Mayer and C.F. Kohn (ed.). *Readings in Urban Geography*, Central Book Depot, Allahabad, p. 287.

Desai A. (1990), Changing Dimensions of Urban Growth, *The Deccan Geographer*, Vol. XXVIII No. 1, Secunderabad, p. 575.

Ibid, p. 576.

Dickinson R.E. (1964), *City Region and Regionalism*, Routledge and Kegan Paul, London, p. 38.

Ibid, p. 167.

Dutt A. K and R. Amin (1986), Toward A Typology of South Asian Cities, *The National Geographical Journal of India*, Vol. 32, Part 1, Varanasi, p. 34.

Garner B.J. (1969), Models of Urban Geography and Settlement Location, in R.J. Chorley and P. Haggett (ed.) *Socio-Economic Models in Geography*, Methuen and Co. London, p. 303.

Ibid, p. 335.

Gautam N.C. (1975), Urban Land use Study of Bikaner city, *The Deccan Geographer,* Vol. XIII, No. 182, Secunderabad, p. 211.

Harris C.D. and E.L. Ullman (1967), The Nature of Cities, in H.M. Mayer and C.F. Kohn (ed.), *Readings in Urban Geography,* Central Book Depot, Allahabad, p. 277.

Harvey D. (1975), Social Processes and Spatial Form: An Analysis of the Conceptual Problems of Urban Planning, in E. Jones (ed.) *Readings in Social Geography*, Oxford University Press, London, p. 290.

Ibid, p. 285.

Jain N.G. (1978), Nagpur Slums: A Problem in Residential Land use, *The Deccan Geographer,* Vol. XX, No. 2, Secunderabad, p. 422.

Johnson B.L.C. (1980), *India-Resources and Development*, Arnold-Heineman Publishers (India), New Delhi, p. 178.

Joshi. S. (1987), The Spatial Pattern of Drug-Retailing in Pune, *The National Geographical Journal of India,* Vol. 33, Part 1, Varanasi, p. 39.

Khatu. K.K. and A.R. D' Souza (1982), Urban Land Use Strategy in Welfare State, *The Deccan Geographer,* Vol. XX, No. 2, Secunderabad, p. 179.

Ibid, p. 178.

Mayer. H.M. and C.F. Kohn (1967), Geography and Urbanism, in H.M. Mayer and C.F. Kohn (ed.) *Readings in Urban Geography,* Central Book Depot, Allahabad, p. 5.

Mc. Gee T.C. (1975), An Aspect of Urbanisation in Southeast Asia: The Process of Cityward Migration in E. Jones (ed.), *Readings in Social Geography*, Oxford University Press London, p. 224.

Mishra R.P. (1978), *Million Cities of India*, Concept Publishing Company, New Delhi, p. 11.

Shafer T.W. (1977), *Urban Growth and Economics*, Reston Publishing Company, Reston, Virginia, p. 133.

Ibid, p. 139.

Sethi M. & S. Pande (1987), Urban Land use changes, A Case Study of Union Territory of Delhi, *The National Geographical Journal of India,* Vol 33, Part I, Varanasi, p. 33.

Tolley R. & B. Turton (1995), *Transport-System, Policy and Planning: A Geographical Approach*, Longman, Harlow, p. 189.

8

SPATIAL EFFICIENCY IN AGRICULTURAL LAND USE

Agriculture is an economic activity which is essentially rural, needing large areas of land for utilization. Land is put to use for agricultural production just as it is used for urban functions in urban areas. Like urban land use, agricultural land use also has spatial significance as it occupies space and evolves a locational pattern. It has, therefore, become a favourite topic of study for geographers. Sir. L.D. Stamp surveyed the entire agricultural land of Great Britain to produce a series of elaborate land use maps. His work won a wide acclaim. Von Thunen built a classical model on agricultural land use pattern which is still useful for several agricultural studies.

Agricultural land use evolves a cropping pattern, suitable to the land and the region. It is, nevertheless, thoroughly influenced by natural conditions of soil and climate and it reflects adaptation with them as efficiently as possible (Singh, 1982, p. 165). The given set of conditions in which the farmers work includes not only natural factors but also the methods of agriculture, irrigation, technological innovations, land rents and other socio-economic factors. Singh and Rajpati Ram (1986, p. 127) find that the basic contributing factor to low production efficiency is the little application of modern technology in the form of various improved production inputs. The part played by agricultural innovations can never be overemphasized. Optimization of agricultural land use is highly related to the innovations in production-distribution techniques viz, irrigation, marketing, fertilizing etc. which leads to efficiency. It is imperative to understand the relationship between the land use pattern and the farmer's resources, abilities, decisions, and his adoption of technology and adaptation to price levels.

The significant characteristic feature of agriculture is its great diversity of practice, products and organization (Pragathi Ramanaiah, 1999, p. 11). As a result, there are immense regional disparities in agricultural land use patterns and in the level of agricultural development. In the agrarian countries like India, where agriculture is the main-stay of the economy, the study of spatial organization of agricultural land use, its spatial variations and its efficiency is of practical significance for suitable agricultural policies for development.

Agricultural land use pattern is interesting from spatial efficiency point of view. It reflects the attempt of the farmer to derive the best use of the land while the distance to market or the transportation costs are minimized. For, a basic premise of traditional economic theory is that the motivating force behind men's actions is the attempt to maximize income, (Found, 1971, p. 1). The agricultural land use pattern that evolves over a great number of years is an evidence of as much efficiency in adaptation to the available conditions as possible on the part of the farmer by trial and error. His activity is controlled by: (i) the use of resources, (ii) the exchange of goods even in so called subsistence economy, and (iii) his decisions of farming system, methods of production, transportation and technology. Ibery (1985, p. 3) opines that the spatial structure of agriculture is the culmination of farmers decisions of how to derive the best use of the land which they control.

Agricultural Landuse Patterns

Agricultural land use is a very complex process which is governed not only by several physical, economic, socio-cultural, technological and administrative factors but also perception of the farmer which influences his decision-making. Morgan and Munton (1971, p. 42) also find his perception substantially influential. The pattern of land use evolves over a period of several years in response to the dynamism of all these factors.

Above all, the locational aspect is primary. For, the distribution of crops in the field creates the pattern. Agricultural land is fixed in location and has an inherent quality. The farmer has to decide which crop and cropping system has to be adopted within the given set so that maximum benefit is derived. The moot question is to determine the use of the land for the right crop. In principle, this problem is identical to that of determining any kind of land use or location of any activity for that matter.

Locational considerations are important for efficiency but they merit particular attention where commercialization is concerned. There again is a difference in subsistence and commercial agriculture as a spatial system of operations. The locational decisions for subsistence crops may be different from commercial crops. A large part of agriculture in the world is subsistence type and a large portion of the world's food output never enters commercial channels (Ibery, 1985, p. 3). India is predominantly in subsistence farming, but even then the commercial or the cash crops are taken. The plantation crops such as tea, coffee, rubber, spices etc. which are totally commercial in nature and quite sizeable in Indian farming form a class of their own. It is a well-organized, labour intensive and most systematic form of cultivation like the tea estates (Barah, 1998, p. 12).

The subsistence farming is intensive. In densely populated countries of the South and South East Asia double cropping and multiple cropping are usual. The agricultural land is hardly left fallow unless necessary. Even the hill sides are not left out. A peculiar type of terrace farming is done on the hill sides. A number of crops are grown on small individual farms as staple foods and some as cash crops to earn money, as suits the soil and climate, particularly the rainfall. As against it, the farming in the western and the developed countries is mostly commercial and extensive on large farms. Rotation of crops and keeping the land fallow are regularly practiced. Modern techniques and innovations are employed to improve production efficiency.

Models of Agricultural Landuse

Models of agricultural land use are constructed to generalize upon the characteristics of land use pattern and to seek the trends and tendencies in the farmer's decision-making.

Most of the earlier models of agricultural land use were economic in nature, concerned with the notion of profit maximization for the farmer. They treated agriculture as a purely economic activity and the farmer as an economic man. They applied economic principles to find out optimacy in land use. Optimum land use patterns rest on the assumption that farmers are rational, have complete knowledge and an equal ability to use this knowledge (Ibery, 1985, p. 7). Since the best or optimum use of land was the aim of such models, they were called optimizer models and they were deterministic. David Ricardo, for example, produced a highly deterministic model of land rents based on the economic principle of comparative advantage. Von Thunen's ideas followed from and developed over those of Ricardo (Tarrant, 1974, p. 21).

The earlier models are well represented by Von Thunen's model. No agricultural land use studies could proceed without a reference to this classical and well-known model which was based on an econometric analysis of his own farm. Von Thunen was a farmer with a prosperous estate near Rostok, and he made observations of general agricultural practices and behaviours of farmers in his locality. The main aim of his theory was to show how and why agricultural land use varies with distance from the market (K. Siddartha, 2000, p. 79). For building of his model, he assumed an isolated state, one central city as the sole market, a uniform plain surrounding the city, uniform mode of transport and land rent related to transportation cost over distance from the market. The prime factor deciding the locational rent was distance from market and the quality or the productivity of land was held constant. Land rent or locational rent was, then, expressed, involving the function of distance, as:

L = E (p-a)-Efk where,
L = Locational rent per unit of land,
E = The yield per unit of land,
P = The market price per unit of commodity,
a = The production cost per unit of commodity,
f = The transport rate per unit of distance, and
k = Distance.

Thunen formed seven land-use rings around the central city in a logical sequence according to the movement input. (Fig. 8.1) The most valuable land use of market gardening and milk production occupies the nearest ring from the centre of the highest rent and so on. These land-use zones were arranged in a concentric circle fashion since one type of land use was homogenously away from the centre. Thus, the model yields a simple circular minimum movement solution with the most efficient shape of the land use zones. The zoning was based on the locational rent according to the distance from the central city. Therefore, the circular zones became increasingly broader from the centre, depending upon tapering fright rates in transportation. Haggett (1968, p. 167) comments, as the general level of transport costs fall, the relative mobility of products increases and we should expect to find Thunen's rings growing wider. This does not destroy the zonal pattern of production. What is interesting from the geographer's point of view is the strong distance element that the model bears forth as affecting the economic activity of agriculture (Found, 1971, p. 57).

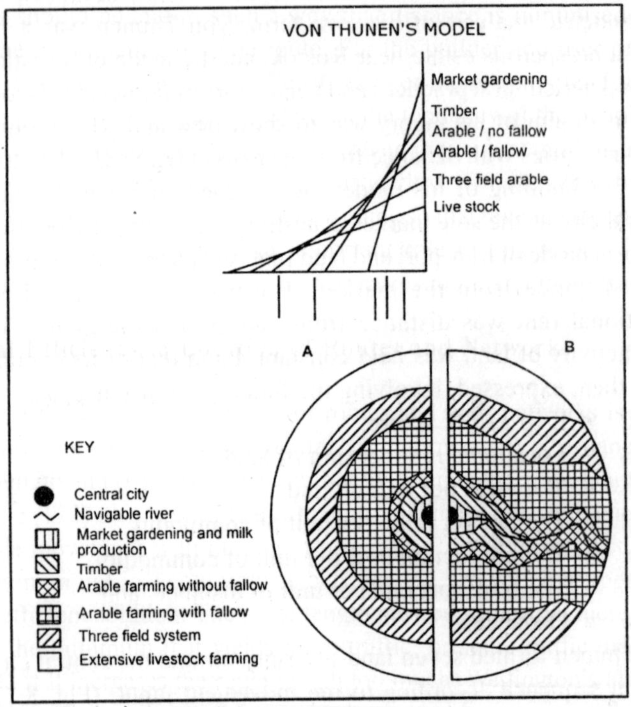

Fig. 8.1: Von Thunen's Model

The standard Thunenian model has been criticized for its simplifying assumptions and its high empirical and deterministic content. Thunen himself revised some of these assumptions accommodating areal differences in productivity, alternative markets and different transport media such as a river, by which the symmetry of the original pattern gave way to an irregular mosaic (Fig. 8.1). As with most approaches to models, Von Thunen first derived a simple structure and later relaxed some of the simplifying assumptions, adding complexity and also bringing the model nearer reality (Tarrant, 1974, p. 26). However, as Harvey (1966) points out, it is a partial equilibrium model. Henshall (1967) also calls it a model of partial equilibrium. It does not take into account the influence of non-economic factors.

Other locational models are derived by analysing comparative advantage of producers, factors of production and consumers at a series of discrete points to explain difference in production at various points. These models are of two basic types: input-output models and spatial equilibrium models (Harvey, 1966, pp. 365-67). Morrill and Garrison (1974) projected trade patterns in wheat and flour in a five-region system of the US using a spatial price equilibrium model.

These models based on economic determinism were gradually replaced by more probabilistic models of farmer's perception and behaviour, thought to be more realistic. The geographer's approach, thus, changed over time from physical determinism to economic determinism and then to farmer's behaviour and decision-making. The objective of the behavioural approach was clear to reject the notion of economic man and replace it with a model that is closer to reality (Ibery, 1985, p. 27). These models are of satisficer kind rather than optimiser. Sub-optimal behaviour on the part of the farmers to achieve some satisfactory return, not necessarily the best, is examined by these satisficer models. There are a number of factors such as choice of fertilizers or risk-taking in which farmer's decision-making plays an important part on the satisficer level. Most prominent among these satisficer models are game theoretic models dealing with probabilities of farmer's decision. The basic game theory by Von Neumann and Morgenstern appeared in 1944 but since then it has been shown to have wide application. It was developed to deal with the problem of optimizing decisions in the face of imperfect knowledge (Henshall, 1967, p. 448).

Diffusion of innovations was another interesting model, based on simulation technique. Bryce and Cross Studied diffusion of hybrid seed corn in Iowa in 1943. But Hagerstrand was the first geographer to develop a model to describe the diffusion of an innovation to the farmers in Sweden in 1952. Hagerstrand (1967, p. 7) himself observed that the spatial order in the adoption of innovations is very often so striking that it is tempting to try to create theoretical models which simulate the process and eventually make certain predictions achievable. Further work in this field has been done by J. Wolpert and R.S. Yuill.

However, it is now becoming clear that the satisficer models are also failing to explain the observed world adequately. Found (1971) pointed out three factors as an obstacle to behavioural theory: (i) that the universal application of the result has not been demonstrated, (ii) that it is difficult to observe human behaviour objectively, and (iii) that human decision-making is a very complex process.

Spatial Efficiency in Agricultural Landuse

The patterns of farmer's behaviour which are empirically established, have to be generalized and interpreted in the light of basic human motives and tendencies. One of the basic motives that controls the economic behaviour of the farmers in agricultural land use is the highest returns and the tendency is to make an efficient use of land to achieve it. In other words, the maximization of land use with minimization of distance to market or the principle of spatial efficiency underlies the pattern of agricultural land use that emerges in the end. This has much in common with industrial location models of the maximum profit type.

Agricultural efficiency is a common concept in agricultural studies and it is measured on the basis of crop yield and cropland. It is being more and more employed to determine agricultural potentiality for a crop of a place or according to R.B. Mandal (1982, p. 342), it connotes the productivity of a unit of land for a particular crop. It is computed by the formula:

$$Ei = \frac{r1c1 + r2c2 + \dots rn\ cn}{c1 + c2 + \dots cn}$$

Where,

Ei = Agricultural efficiency of ith place,

r = Ranking of yield of crop, and

c = Cropland share in per cent.

Bhaskar and others (1987, p. 154) have the same opinion that agricultural efficiency is a scientific devise to study the inherent fertility and capability of land. They give the formula to compute it as follows:

$$\text{A.E.} = \frac{\text{Di}(\text{X1} + \text{X2} + \text{X3} + \ldots \text{Xn})}{\text{R}\,(\text{X1} + \text{X2} + \text{X3} + \ldots \text{Xn})} \times 100$$

Where,

A.E. = Agricultural efficiency,
Di = Crop yield in ith district,
X1 – Xn = Crops grown in the region,
R = Regional crop yield

Agricultural efficiency measures one of the parameters of spatial efficiency *i.e.* the use of agricultural potentiality that is reflected in crop yield. It has to be combined with the minimization of distance factor to find out spatial efficiency. Thunenian model has a spatial efficiency element as it considers distance and transportation cost from the market in the centre. The land use changes away from the centre in the concentric zones according to the changing land rents based on distance. Thus, distance is considered as an explanation of land use pattern. Though the model does not take into account the productivity of land which also affects the land rents and land use pattern, it really is very close to the principle of spatial efficiency as it proposes the maximization of highly valued land by taking high-valued crop there and so on.

All agricultural systems, from subsistence to highly commercial, are dependent on land along with other factors of production on the one hand and demand for the product on the other. Thus, two sets of economic relationships interact to determine the choice of crop on land: supply-and-demand (Morgan and Munton, 1971, p. 6). The supply-demand-price syndrome is affected by distance. The effect of distance on the pattern of land use has been changed by advancement in transport technology (World Development Report, 1986). Fast means of transportation are reducing distance on a large scale but they cannot write off the relationship between distance and land use. Distance is important in one form or the other. Chisholm (1962, p. 48) has shown that the land use around a farmstead is directly related to the distance between farmhouse and field. In the developing countries, the factor of distance is still pronounced and there are several cases where the land use around the rural settlement is directly related to distance from it (On Africa, Prothero, On India, Ahmad, On Brazil Waibel, cited by Henshall, 1967, p. 445). At the same time, the demand from the market works as a force making the farmers maximise the potentiality of land, and choose the crop accordingly. Thus, the basic postulates of spatial efficiency are never outdated.

The agricultural land use models are now seeking more and more maximum-minimum solution. The land use allocation models using linear programming are maximum-minimum type. Heady and Egbert (1964, pp. 374-86) used linear programming techniques to allocate the production of field crops within 122 producing regions of the United States. Such a maximum-minimum criterion was used by Wald or Dillon and Heady for the best solution of the selection of crops, for maximum return with minimum risk-taking. Morgan and Munton (1971, p. 37) call the game theoretic models as 'maximin' models seeking maximum-minimum solutions. Spatial efficiency model also implies maximum-minimum solution in spatial context. It aims at maximization of land use for a particular crop and minimization of distance or transportation cost of it to the market. It is a spatio-economic model and of an optimizer kind with the belief that the farmer is basically an optimizer in spatial organization of agricultural land use, though he may refrain at the satisfying level.

Spatial efficiency considers maximum utilization of productivity of agricultural land for a particular crop on the one hand and distance from the market or the transportation cost on the other. Productivity of the land, investment of inputs in the production and element of risk are very well reflected in the yield of crop. The low level of yield invariably reveals poor sychronization of crop with any of these things, and failure in maximization factor. The production of crop should be viewed along with a price it can fetch in the market, even though it is not marketed. The total value of the production is compared against distance from the market. Distance is important to know even for subsistence crop because there is some surplus of this crop for sale. No farmer, even a subsistence farmer can operate in isolation and along with subsistence crops, the farmer takes some cash crops for sale to earn money. Therefore, distance or transportation cost can be used as denominator in spatial efficiency equation. In Indian context, transportation cost may be misleading, as the costs are too much varying for different means of transportation and many times the farmers carry the produce in their own bullock carts for which they have to pay in time and not in money.

Distance directly affects land rents, but unlike Thunenian presupposition, land rent is not only a function of distance but also that of the quality or capacity of land. It is highly competitive and sensitive index. But it cannot be used in place of distance since it is not a minimizing factor. So, the formula for computation of spatial efficiency of land use is as follows:

$$SEcri = \frac{Yixv}{dij}$$

Where,

SEcri = Spatial efficiency of a crop at ith place,
Yi = yield of crop at ith place,
v = Value per unit of crop and
dij = distance from ith place to market j

This measures efficiency of existing crop. A new crop can also be simulated by taking expected yield of it instead of current yield for computation. In the above formula, average yield per unit of land can also be used in place of total yield which is a better measure of capacity of land. This measure is quite useful in comparative studies but the optimacy is hard to find. The computation of spatial efficiency can be done with agricultural efficiency also which uses crop yield and per cent share of cropland, if it is known, as:

$$SEcri = \frac{AEcri}{dij}$$

Where,

SEcri = Spatial efficiency of a crop at ith place,
AEcri = Agricultural efficiency of a crop at ith place, and
dij = distance from ith place to market j

Spatial Efficiency of Agricultural Land Use in Vidarbha

Agriculture is the largest economic activity of the people of Vidarbha and so, its economy is predominantly agrarian. Agriculture is traditional subsistence type with a few cash crops like cotton, groundnut, soyabean, fruit and vegetables (Arunachalam, 1967, p. 113). The major food crops are millets (Jawar and Bajra), rice, wheat, and pulses grown in Kharif and Rabi seasons. Irrigation is little—only on about 9 per cent of cultivable area and it is mainly tank or well irrigation. The cropping pattern and crop combination, therefore, depend to a large extent, on the amount of rainfall. Rice needs high amount of water, which is grown in wetter: Eastern Vidarbha and the hills of Melghat in Amravati district. In the rest of the drier Vidarbha cotton and millets (Jowar) is the popular combination.

Fig. 8.2 shows the major crop combination regions. The cropping pattern has evolved over many years of farmers decisions and it is best suited for the region. But as Andrade (1972, p. 321) opined in his project report, the agriculture is loosely-managed and there is a need for planning to ensure optimal economic and social returns. Optimization in agriculture or spatial efficiency will be achieved by improving agricultural efficiency or crop yield per unit of land. B.B. Singh and Rajpati Ram's (1986, p. 134) findings for Azamgarh district are also the same *i.e.* intensifying the cultivation and raising crop yields.

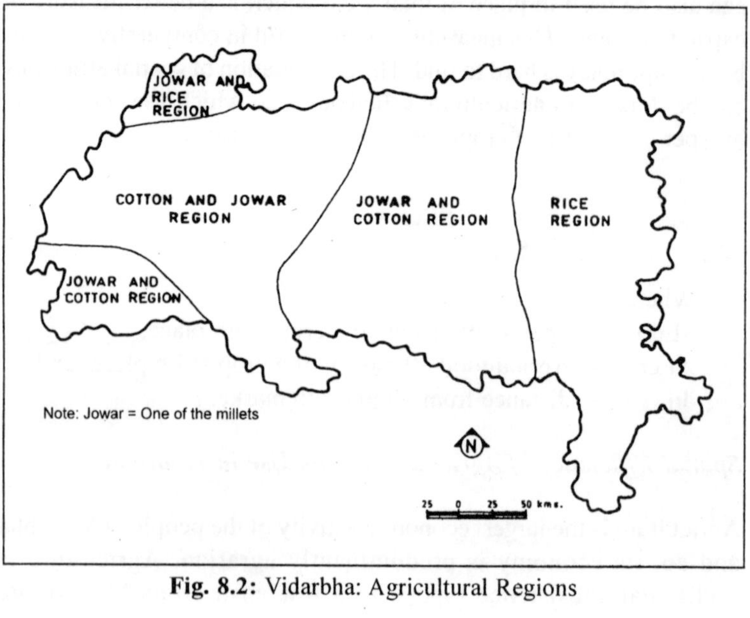

Fig. 8.2: Vidarbha: Agricultural Regions

If the Thunen's model is applied to the crop pattern of Vidarbha, it is found that the model does not at all fit with it. Undoubtedly, the land use zones in the Thunen model are in keeping with the European pattern with rotation and timber production but allowing for Indian conditions, the zoning according to the land rents as in Thunen model, is not discernible. The same type of crop combination but for a few special crops is found everywhere irrespective of the distance from the market. Since the valuable land is not utilized for valuable crops and thus maximization of land use is not achieved, the agricultural land use is inefficient. The returns are not maximized and it perpetuates the poor state of the farmers and inefficiency.

A land use survey in village Parlam in Amravati district of Vidarbha will serve as a case in point. Village Parlam is just 15 kms. to the west of Amravati city, an urban market. Agriculture is the main occupation of the people and out of total 523.27 hect. of area, 478.31 hect. *i.e.* 91.41 per cent area is cultivated. The following Table 8.1 gives the land use in village Parlam.

Table 8.1: Land Use in Village Parlam

Land use	*Cultivable Land*	*Govt. Land*	*Irrigated Land*	*Forest*	*Pasture*	*Total*
Area (hect)	478.31	16.15	13.13	7.94	7.74	523.27
Per cent	91.41	3.09	2.51	1.52	1.47	100.00

Source: Gram Panchayat Parlam.

The cropping pattern is adjusted to the natural set of conditions. Cotton-Red gram-millets are the main kharif (summer) crops and Gram-wheat-vegetables the main rabi (winter) crops. Among these cotton is the main cash crop and then soyabean vegetables and oranges on a small scale. The following Table and Fig.8.3 show the crops in land use.

Table 8.2: Area Under Crops in Village Parlam

Crop	*Cotton*	*Red Gram*	*Millets*	*Gram*	*Wheat*	*Soyabean*	*Vegetables & Oranges*	*Total*
Area (hect)	145	106	86	49	20	30	15	451
Per cent	32.15	23.50	19.07	10.87	4.43	6.65	3.33	100.00

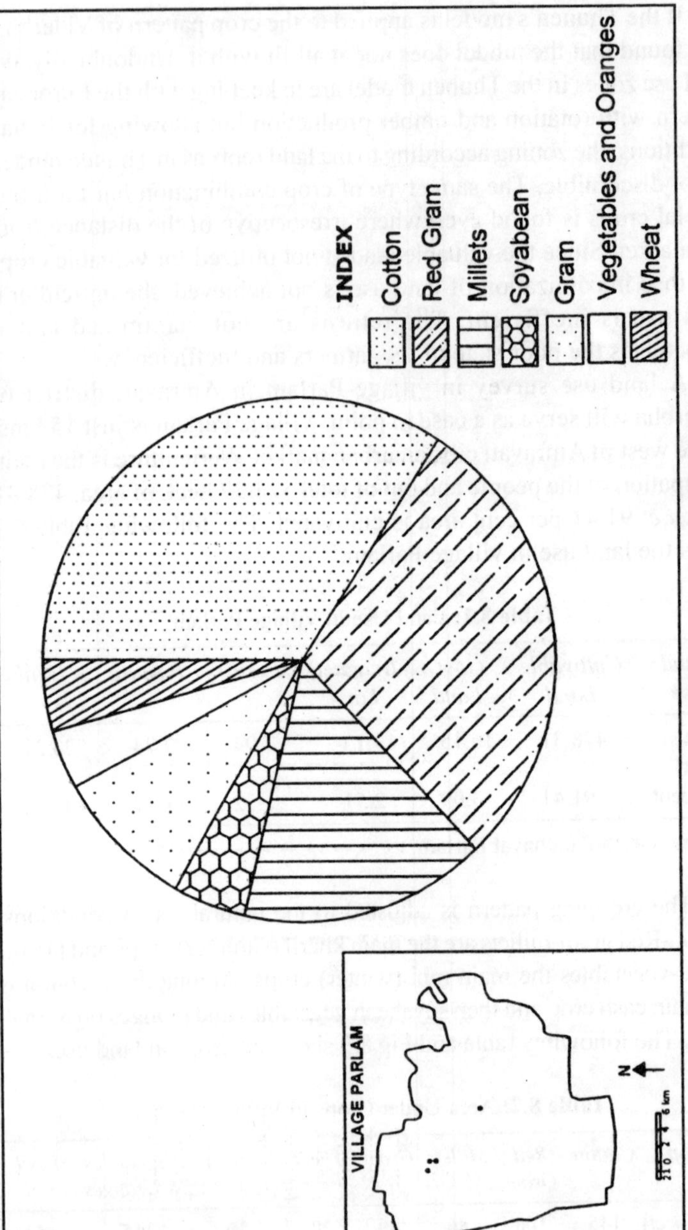

Fig. 8.3: Area Under Crops in Village Parlam 2002-2003

The cropping pattern is just like the general pattern in this part of Vidarbha inspite of Parlam's proximity to market. Cotton, soyabean, vegetables and oranges, the cash crops are sold in Amravati. The village is joined to Amravati by a metalled road and agricultural commodity is carried by the road mostly in farmer's bullock-carts. The cash crops do not have separate zones but are mixed with other crops in the fields.

Had Thunen's model been any guide, Parlam's land would have been largely used for high-valued crops like fruit and vegetables, since the village is very near Amravati. In the nearest zone of the market, the model finds market gardening and milk production on the highly valued land which can also fetch high price and become economically viable and spatially efficient. Agriculture being subsistence in nature, the agricultural land in Village Parlam cannot be wholly devoted to market gardening. But a substantial part of it should have gone to it. This agricultural area fortunately, belongs to the famous orange belt of Vidarbha (Fig. 8.4). The quality and productivity of the land without which the crop cannot sustain in the area is high for orange cultivation and market of Amravati is ready nearby. The oranges, vegetables and even flowers would be a better and efficient combination in the crop pattern which would be beneficial for the farmers.

But actually the per cent share of oranges and vegetables in the crops is very low (3.33%), which means that the land has not been put to efficient use. Probably the reason for it is the lack of irrigation facilities as only 2.5 per cent of cultivated land is irrigated, and market garden crops like vegetables, fruit and flowers need a lot of irrigation water. A.L. Singh and F. Azam (1986, p. 103) have found a high correlation between irrigation and crop output in Western U.P. and suggest irrigation as a very important ingradient for agricultural production. It is more so for garden crops. In addition to irrigation facility, a large sum is also required as investment in inputs for oranges which only the rich farmers can afford. The general economic conditions of the farmers prevent them from optimum utilization of land and they turn to traditional cropping pattern which perpetuates spatial inefficiency. It is because maximization of land use is not done even though the distance to the market is minimum.

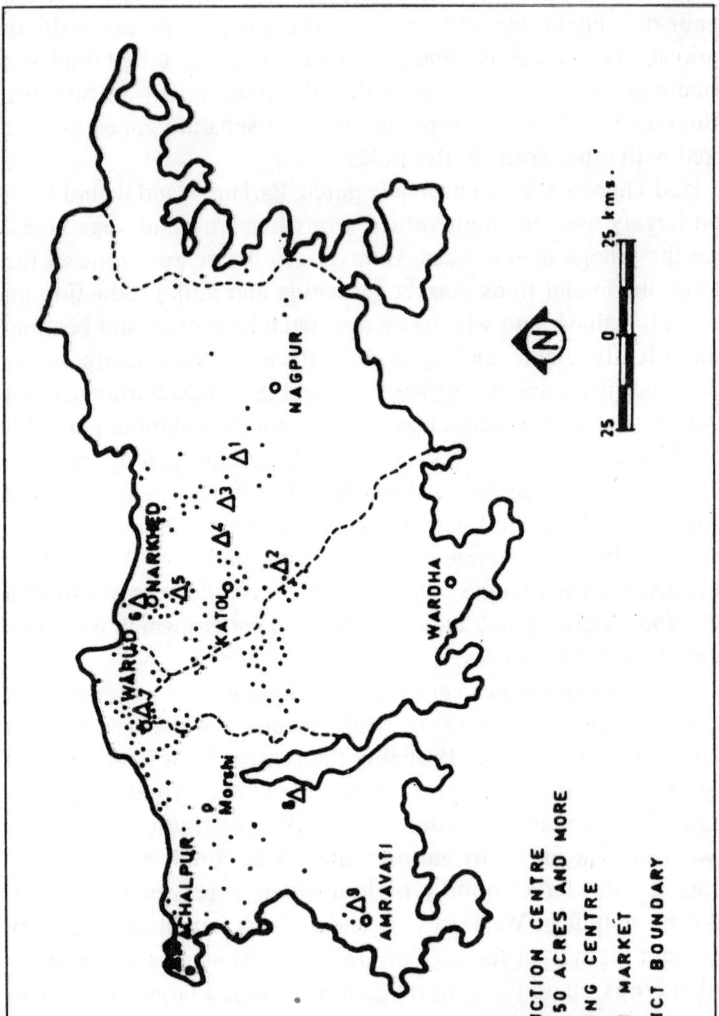

Fig. 8.4: Orange Production Centres in Orange Belt of Vidarbha

Although village Parlam does not grow oranges, the mid-northern part of Vidarbha in which Parlam is included, has specialized in good quality orange production. It comprises parts of Nagpur, Amravati, Wardha and Bhandara districts, covering an area of 16065 kms^2, known as the orange belt. It excels in orange production not only in Vidarbha but also in the whole of India (Fig. 8.4). The terrain conditions like soil, slope and drainage and climate are very suitable for good quality mandarin type of oranges with a super flavour and taste. Thus, the orange belt utilizes the potentiality or capacity of land for orange cultivation. Spatial distribution of orange production and packing centres is shown in Fig. 8.4.

The big markets of oranges are at Nagpur, Amravati, Kalmeshwar, Narkhed, Warud, Morshi, Achalpur, Anjangaon and Wardha. Transportation is important in their marketing. Orange production, tied down to the land and its capacity to bear good quality fruit, has to be maximized in the area which is rather away from the big urban centres. Minimization of distance to market, therefore, has been efficiently achieved by creating orange markets in the producing area itself at Narkhed, Kalmeshwar, Warud, Morshi, Anjangaon and Achalpur, apart from the urban centres of Nagpur, Amravati and Wardha. This is spatially more efficient.

Spatial efficiency in orange production has been measured for three districts in Vidarbha, namely, Nagpur, Amravati and Wardha. A very small part of Bhandara district is also included in the orange belt, but its share in total distribution-production is so negligible that it can be left out. The formula for measurement is:

$$SEo = \frac{YxV}{\underset{i=1, j=1}{dij}}$$

Where,

SEo = spatial efficiency of oranges,

Y = average yield of oranges per unit of land,

V = value of oranges per unit of weight,

d = average distance from i th producing centre to j th market.

Table 8.3: Spatial Efficiency of Orange Production in Vidarbha

District	*Market*	*Average Yield per hect. (m.tons)*	*Value (Rs) per m.ton*	*Value (Rs) per hect.*	*Mean dist. (Kms)*	*SE*
Nagpur	Nagpur, Narbhed Kalmeshwar, Katol	15	2596	38,940	10.5	3708
Amravati	Amravati, Warud Morshi, Achalpur Anjangaon	16	3100	49,600	12.0	4133
Wardha	Wardha	14	2000	28,000	16.7	1677

The best fruit is produced in Warud cell of Amravati district and Katol cell of Nagpur district. The area under oranges is increasing in Achalpur and Anjangaon Tahsil in Amravati district and the fruit fetches better price. Average distance to market includes average distance from producing centre to packing centre and that from packing centre to the market. It is slightly lower in Nagpur district but spatial efficiency is better in Amravati district as the yield per hectare and its value is more in it. Spatial efficiency is the comparison of both the parameters and it does not depend on the distance factor alone. In case of oranges, the yield per hectare and value of fruit per hectare fluctuate much from year to year and therefore efficiency is also liable to change. Spatial efficiency in the orange belt will increase if the area under the fruit and per hectare production of the fruit could be maximized. But that depends upon better irrigation facilities, fertilization and pest control.

The land under the oranges is highly valued and the land rent ranges from Rs. 3,75,000 to Rs. 6,25,000 per hectare according to the accessibility to the metalled road. It is in keeping with the Thunen's tenet that highly valued land is devoted to highly valued crop. But the high-value land does not form a ring around an urban market as he postulates, because in this case, the land rent changes according to the productivity of the land for oranges and not according to distance from market. Von Thunen's model, which does not consider the capacity of land, does not apply everywhere and also fails to explain the spatial efficiency of production.

There are many empirical deviations from Thunen's ideal though his thesis cannot be negated outright. In a case study of agricultural development around the city of Indore in the State of Madhya Pradesh, Y. Joshi (1983, p. 433) found an interesting fact that there was definitely not a ring pattern around the city of Indore as Thunen envisaged. Rather it is axial along the main road and sectoral in the area with better irrigation facilities. But it is true that development and commercialization of agriculture is generally taking place near the city as the market is nearby. Now-a-days, there is a growing tendency of truck farming at the rural-urban fringe which has a ready market at hand. The land rent is high at the rural-urban fringe and it is profitable and spatially efficient to use it for highly valued and perishable garden crops of vegetables, fruits and flowers.

Conclusion

Efficiency in agricultural land use is crucial in a predominantly agrarian country like India. Agricultural practices and land use patterns have evolved over a long period of time through trial and error and are guided in the process by physical and spatio-economic factors. The empirical land use studies reveal that the traditional crop patterns do not prove to be quite efficient in use of land in the changing conditions. The agricultural land use would be profitable if it is spatially efficient. There is a trend to make it efficient and profitable, especially in respect of commercial crops such as oranges in the orange belt of Vidarbha or truck farming near the rural urban fringe. However, maximization of crop production to withstand the land rent is necessary for the farmer to be efficient and profit making. The principle of spatial efficiency can prove to be a guide to the farmers in their choice of crop and the cropping pattern. More studies from spatial efficiency point of view are required to help the farmers in their decisions to optimize their land use.

Many models have been constructed to explain and optimize the agricultural land use and the most renowned among them is that of Von Thunen's which is very near the efficiency consideration. First, these models were deterministic but gradually they gave way to more probabilistic and behavioural models which were of satisficer kind. Now, the trend is again towards optimising models in which linear programming is applied and maximin factors are employed. This again implies the concept of spatial efficiency. The spatial efficiency model is able to compare efficiencies of existing land use in crops of different areas, and also to simulate the efficiency in production of a particular crop in an area.

The age-old choice of crops and methods of farming are an outcome of the farmer's decisions and any change in them for better is again due to the farmer's decision making. His decision-making is influenced a great deal by his information and perception of realities. Geographers are now believing that perception has a key role to play not only in the farmer's behaviour but also in the general spatial behaviour of man. Perception has, thus, come into being as a new area of geographical study and it forms the topic of the next chapter.

REFERENCES

Andrade P. (1972), Pilot Research Projects in Growth Centres, in A.K. Sen (ed.), *Reading on Micro-level Planning and Rural Growth Centres*, National Institute of Community Development, Hyderabad, p. 321.

Arunachalam B. (1967), *Maharashtra,* A.R. Seth and Co. Ltd., Bombay, p. 113.

Barah. B. (1998), Size of the Tea Estate and Productivity of Tea: A Case Study, *The Deccan Geographer,* Vol. 36, No. 1, Pune, p. 12.

Bhaskar. K.S, Sohan Lal, O. Challa and S.H. Madhavi (1987), Agricultural Efficiency of Vidarbha Region, (Maharashtra), *The National Geographical Journal of India,* Vol. 33, Part 2, Varanasi, p. 154.

Chisholm M. (1968), *Geography and Economics*, G. Bells and Sons Ltd, London, p. 48.

Found W.C. (1971), *A Theoretical Approach to Rural Land use Patterns*, Edward Arnold, London, p. 1.

Ibid, p. 57.

Hagerstrand T. (1967), On Monte Carlo Simulation of Diffusion, in W.L. Garrison and D.F. Marble (ed.), *Quantitive Geography, Part I: Economic and Cultural Topics*, North-western University Press, Evanston, Illinois, p. 7.

Haggett P. (1968), *Locational Analysis in Human Geography,* Edward Arnold (Publishers) Ltd. London, p. 167.

Harvey D.W. (1966), Theoretical Concepts and the Analysis of Agricultural land-use Patterns in Geography, *Annals of the Association of American Geographers*, 56, pp. 365-67.

Heady E.O. and A.C. Egbert (1964), Regional Programming of Efficient Agricultural Production Patterns, *Econometrica,* 32, pp. 374-86.

Henshall J.D. (1969), Models of Agricultural Activity, in R.J. Chorley and P. Haggett (ed.), *Socio-Economic Models in Geography,* Methunen & Co, London, p. 448.

Ibid, p. 445.

Ibery B. (1985), *Agricultural Geography: A Social and Economic Analysis,* Oxford University Press, Oxford, p. 3.

Ibid, p. 7.

Ibid, p. 27.

Joshi Y.G. (1983), Agricultural Development Around a City: A Case Study of Indore Tahsil, M.P., *The Deccan Geographer,* Vol. XXI, No. 3 Secunderabad, p. 433.

Ibid, p. 437.

Mandal R.B. (1982), *Statistics for Geographers and Social Scientists,* Concept Publishing Company, New Delhi, p. 342.

Morgan W.B. and R.J.C. Munton (1971), *Agricultural Geography,* Methuen and Co. Ltd., London, p. 42.

Ibid, p. 6.

Ibid, p. 37.

Pragathi P.S. and Y.V. Ramanaiah (1999), *Agricultural Regionalization*: A Theoretical Perspective and its Applied Significance, *The Deccan Geographer,* Vol. 37, No. 2, Pune, p. 11.

Singh A.L. and F. Azam (1986), Correlation of Irrigation and Crop Output in Western U.P., *The National Geographical Journal of India,* Vol. 32, Part 2, Varanasi, p. 103.

Singh B.B. and Rajpati Ram (1986), Food Production System and Efficiency in Ajamgarh District, *The National Geographical Journal of India,* Vol. 32, Part 2, Varanasi, p. 127.

Singh M. (1982), Agricultural Characteristics of lower Ganga-Ghagra Doab, *The Deccan Geographer,* Vol. 20, No. 1, Secunderabad, p. 165.

Ibid, p. 134.

Siddartha K. (2000), *Economic Geography: Theories, Processes and Patterns,* Kisalaya Publication Pvt. Ltd., New Delhi, p. 79.

Tarrant J.R. (1974), *Agricultural Geography,* David and Charles Ltd. Newton Abbot, p. 21.

Ibid, p. 26.

9

PERCEPTION OF SPATIAL EFFICIENCY

Space perception as related to spatial behaviour of man is comparatively new approach to geographical understanding. Formerly the responses of man to environmental or spatial phenomena were analysed without taking into account the process that goes before the behavioural response of man—the process of perception, that becomes an intervening variable. Burton (Downs, 1970, p. 67) referred to perception as coming to merit a place alongside the quantitative revolution in terms of significant new viewpoint.

Geographic space perception is a part of a trend in modern geography known as the 'behavioural revolution' following in the wake of the quantitative revolution. It touches upon psychology for the fundamental concepts of perception, but as Gould (1966) argues, psychological literature does not have much to offer on perception of space, which is a geographical field. Says H.C. Brookfield (1970, p. 53), 'There is a growing body of geographers who see in perception a concept which can unify the several strands of human geography'. A very useful discussion of the literature on perception is given by Saarinen (1966) who is one among the geographers to have made empirical use of psychological methods.

Perception studies concentrate on the cognitive understanding that man has of his space and the way in which this knowledge is stored and organized in his mind. Thus, perception is the way in which a man sees his space or environment and makes a subjective evaluation of it. It forms a cognitive image of the real world in his mind. His spatial behaviour is a function of the image of the world that he carries and not of the real world itself. Perception forms a crucial link between the real world and man's behavioural response to it.

A subtle process goes on in the mental set-up of man as the real world space is perceived before he responds to it. Fig. 9.1 is a simple schematic representation of perceptual stages that a man goes through. An image formation precedes his decision-making leading to behaviour and the image is formed through his value system that includes his attitudes, abilities, beliefs, and tendencies. So, the same space may be perceived differently by different persons. Gollege and others (1987, pp. 3-13) explored the cognition of adults, children and mentally challenged persons and found them quite different. The perceived space is, therefore, monistic and very complex and there is the dual problem of the nature and quality of information available to the decision maker on the one hand and his ability and attitude to make use of it on the other. Pred (1967) took into account this dual problem of the perceived environment and prepared a 'behavioural matrix' to find out the probabilities of decisions. Man's behavioural decisions and the resulting spatial behavioural patterns, thus, depend on his perception of the real world rather than the real world itself.

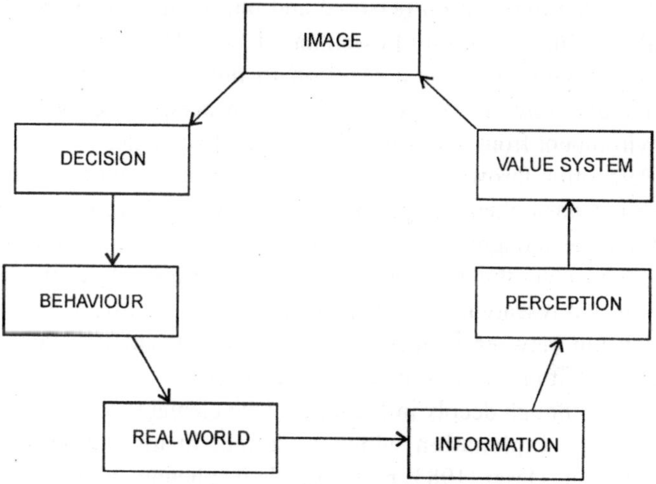

Fig. 9.1: A Process of Perception

Man always has to interact with space and spatial dimensions. Invariably, therefore, his spatial images are related to location, distance, direction and area, which are basic to his spatial behaviour. He conceptualizes, for example, in a route image, the sequence point, salient landmarks en route and the destination, all topological in character rather than strictly metric (Gollege, and others, 1987, p. 1). After non-

Euclidian geometries, the duality of physical and possible space was recognized. The possible space fits the concept of space perception. Since his spatial behaviour is controlled by his perception of space, it goes without saying that his endeavour for spatial efficiency is influenced by it. The study of spatial efficiency is not meaningful without due recognition of the underpinnings of human perception.

Perception of Man/Environment Relationship

The earth's space is full of environment (Kirk, 1975, p. 97) with which man co-exists and he develops so many interrelationships and associations with his environment. His space perception or its equivalent environmental perception plays an important part in these interrelationships and their patterns. Kirk (1951) called it 'behavioural environment' or an internal environment that conditions behaviour. Much of this thinking on personal and universal perception of the world was summed up by Lowenthal (1961). This behaviouralism coming to geography as a reaction to positivist approach with considerable force in 1960s provides a new perspective for looking at environment from man's own inner cognitive angle.

Perception throws a new light on man/environment relationship which is a basic theme in geography and the influence of which on geographical thought is indisputable. Haggett (1975, p. 11) among others considers this relationship as the major aim of geographical analysis and research. It develops into systems analysis as two-way relationship between man and environment with input, output and feed-back within the whole set-up of eco-system. However, human behaviour, though deeply influenced by the elements of eco-system, needs more explanation which is sought in his perception of environment. Desai (1983, p. 333) calls it 'phenomenology' or the man's perception of environmental phenomena affecting his decisions to deal with his environment. This concept focuses upon the awareness and use of his environment through his own evaluation. Thus, in the phenomenological framework, attention is given to the analysis of his subjective environment as it is sensed by him as a human being.

Man, being a part and parcel of the environment and using it for his livelihood, devises many ways to utilize it gainfully. The whole of cultural environment that he has created interacting with natural environment, is an expression of his ability to modify it to his own good. To use it, he perceives it first, forms its mental image and then decides his action. His mental images with spatial dimensions have relative locations which may get distorted and may not exactly correlate with the real space. There are over or under estimation of perceived distances and discrepancies in the mental directions. Several empirical studies have been carried out on cognitive maps of cities. Many examples of such cognitive maps appear in the books by Downs and Stea (1977) and by Gould and White (1986), and they reflect the images that people have of cities in their own cultural context. Gollege et. al. (1987, p. 7) have also produced the cognitive maps by different persons, of the city of Columbus in Ohio, with a distinct downtown, a freeway system and sets of cues, well-known to its inhabitants. The maps are much distorted. However, why specific disturbances occur in mental maps remains mostly unanswered. The authors (*Ibid.* p. 14) conclude that finding out how people use spatial cues and concepts in their everyday lives gives them the hope of being able to understand more completely their activity patterns and decision processes. This knowledge should lead to better theories, models and predictions of different types of human spatial behaviour.

Like cognitive mapping, empirical surveys in perception of people highlight some interesting facts which explain their behavioural pattern. Kavita Ranga (1990, pp. 257-264) carried out an elaborate survey of the Primary Health Centre staff in rural Haryana. To her dismay and against the common notion that the doctors and other staff are reluctant to serve the rural areas, she came to know that they were interested to work in villages but for the absence of basic facilities and social amenities there. Such empirical studies are needed to identify the real problems that lie athwart the implementation of the environmental planning and development programmes. It is easier to provide guidelines for planning if the subjective concepts of residents are known and fully understood. The perception studies of man/ environment relationships have a long a way to go.

Spatial Efficiency in Man/Environment Relationship

Man is an active agent in his interaction with environment. His behaviour is conscious and cognitive in relation to it (Gollege, 1980, pp. 14-15). There is an endeavour on his part to make his spatial relationship with environment as fruitful as possible. Firey (1975, p. 138) calls it 'economic ecology'. It means, according to Swaminathan (1980) maximizing the environmental benefits and minimizing the risk. There is also an endeavour on his part to minimize effort, especially in relation to distance friction in spatial sense and thus to be spatially efficient. This basic tendency of man rules supreme in his spatial behaviour in relation to environment. However, his endeavour is conditioned by his ability and know-how. The resulting behavioural patterns, therefore, differ from region to region not only with differing environmental conditions but also with different responses of man to them. Hence, it may happen that near about similar environment gives rise to different responses and different spatial behaviours. Thus, primitive people were deprived of their land because more advanced people perceived how to master it more effectively and efficiently. Primitive people fell short in spatial efficiency in their milieu due to their lack of knowledge and techniques to make the most of the environmental potentialities.

There are a number of situations of man/environment relationship which could be analysed in terms of spatial efficiency. Before multipurpose river dam project is taken up, it is necessary to determine if it is spatially efficient as well as economically viable. It should have a site at a minimum possible distance from the maximum possible number of people to be benefited. At the same time, the displacement of the people over the space due to the project should be minimum, *i.e.* not so much as to nullify its advantages. In a case of land reclamation and its distribution for urban and agricultural purposes, Maurice Yeates (1974, p. 240) has applied a linear programme model and arrived at a primal-dual solution of 4000 hects to be devoted to urban uses and 6000 hects to agricultural uses. He has shown how maximization of land use over and above the cost of land reclamation would be achieved. It is a spatially efficient solution leading to optimum utilization of land.

Perception of Spatial Efficiency

Spatial efficiency depends, to a large extent, on how man perceives it. Man has to perceive spatial efficiency before he strives for it. Like the subtle process of perception of environment that goes in between environment and spatial behaviour of man, his efficient behaviour follows the perception of it. Man first perceives the spatial or environmental potentialities that he has to maximize by his activity and distance that he has to minimize and then endeavours to attain it. Thus, there is a conscious effort on the part of man for spatial efficiency which he has to perceive as rightly as possible (Fig. 9.2). There are however, different levels of attainment of spatial efficiency as there are differences in the perception of it as well as in the capacities from person to person. If the environment is not correctly perceived, the endeavour for spatial efficiency is liable to go wrong. Saving the distance, for example, cannot be appropriately achieved if the distance is over or under estimated. So also, the changing conditions of environment have to be rightly perceived for spatial efficiency. The Indian agricultural land use is still mostly traditional and it is adapted to the environmental conditions over the years. But some of the conditions change. When the farmer does not perceive them rightly he loses spatial efficiency.

Spatial efficiency is also not fully attained if only one of the maximum-minimum factors is paid attention to at the cost of the other. In Indian cities, for example, minimum distance to the goods and services is sought but maximization of land use is ignored and the level of efficiency goes down.

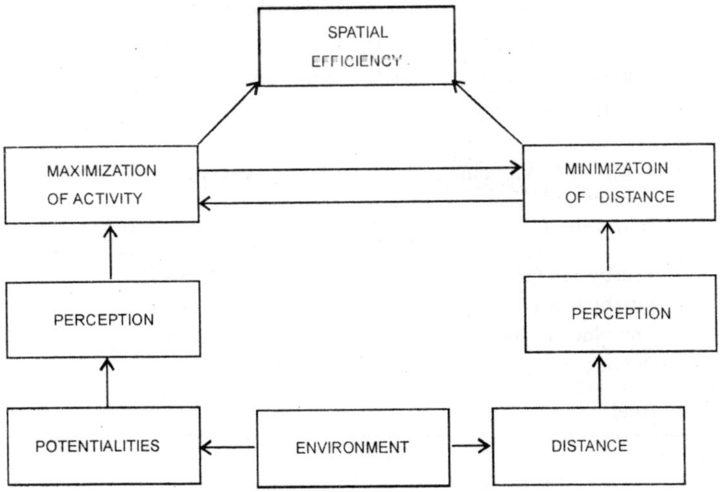

Fig. 9.2: A Schematic Representation of Perception of Spatial Efficiency

There is a growing interest of geographers in environmental perception of man. Nag (1983, p. 219) comments that man and environment paradigm includes the notion that individuals and groups relate to their environment through their perception or cognition of it. His perceived environment includes the factors which he has to maximize and minimize for spatial efficiency. It follows that the studies of environmental perception should be extended to cover perception of spatial efficiency also.

Perception of Spatial Efficiency in Movement of Tribal Youths in Gadchiroli District of Vidarbha

In a sample survey of perception of movement among tribal youths in the age group of 15-35 years in 43 villages of Gadchiroli district of Vidarbha, an attempt was made to analyse it on the basis of the principle of spatial efficiency. Gadchiroli is a forested and predominantly tribal area in the easternmost part of Vidarbha. It is backward with bare facilities of transport and communication available to the tribals. Though the road density has now increased to nearly 35 kms. per km^2 besides a very short narrow gauge railway passing through the district and some waterways to cross in small canoes, tribals live in this region quite isolated, in their own microcosm, their same traditional primitive world. Both the cause and consequence of it is their least mobility, which is restricted upto a nearby village for marketing, fair, meeting the relatives and schooling. The following Table shows movement of tribal youths.

Table 10.1: Movement of Tribal Youths in Gadchiroli District

Destination	*% of Total Youths Moving*
Nearby village	88.2
Nearby town	11.2
Any place in India	0.6
Outside India	–
Total	100.0

As regards their perception of movement, most of the youths were not inclined to travel long distances. They disliked and avoided journeys. One of the reasons was that the travelling facilities were poor. The infrequent bus, the bullock cart and the canoe in the river made up all the vehicles that were available for travel and nearly 32.5 per cent of travelling youths had to go on foot from one place to the other, so, they kept the travel minimum to only those occasions when journey was a must. They did not perceive to maximize movement, while minimizing distance. To cut the movement itself was not the proper way of spatial efficiency. But no single person wanted to go places, to see the world, to learn new things or even to move away for jobs.

There was a kind of vicious circle. Inefficiency in movement gave rise to isolation, both spatial and social. It, in turn, badly affected the perception of spatial efficiency in movement of tribal youths. To perceive spatial efficiency in movement rightly was the way to cut this vicious circle and to promote developmental activities, because the need for development should come from within the society and not to be imposed by an external agency (Desai, 1983, p. 335).

Conclusion

Perception is a key factor between man and his behaviour. It needs to be studied to understand man's spatial behavioural patterns better. However, much has to be done to refine methods and techniques in such studies. Perception is an intricate and subtle process and still in its infancy as a topic of study. It really needs a thrust, because it underlies all human spatial behaviour. To understand spatial efficiency in locational and land use activities, an insight in perception of spatial efficiency is useful. It was found in the study of perception of spatial efficiency in movement among tribal youths of Gadchiroli district that their movement was on a very low key, leading to their isolation and backwardness as they did not perceive the spatial efficiency in movement in the right way. More and deeper studies on perception and model building in this field would do much to develop geographical understanding of human behaviour.

REFERENCES

Brookfield H.C. (1971), On Environment as Perceived, in Board et al (ed.), *Progress in Geography,* Vol. 1. Edward Arnold, London, p. 53.

Desai. A. (1983), The Framework of Man/Environment Relationship: Changing Philosophy in Geography, *The Deccan Geographer*, Vol. XXI, No. 2, Secunderabad, p. 333.

Ibid, p. 335.

Downs R.M. (1970), Geographic Sapce Perception: Past Approaches and Future Prospects, in C Board et. al (ed.), *Progress in Geography,* Vol. 2, Edward Arnold, London, p. 67.

Firey W. (1975), Sentiments and Symbolism as Ecological Variables, in E. Jones (ed.), *Reading in Social Geography,* Oxford University Press, London, p. 138.

Gollege R.G. (1980), A Behavioural View of Mobility and Migration Research, *The Professional Geographer,* Vol. 32, No. 1, pp. 14-15.

Gollege R.G., N. Gale and G.D. Richardson (1987), Cognitive Maps of Cities (II): Studies of Selected Population, *The National Geographical Journal of India,* Vol. 33, Part 1, Varanasi, p. 1.

Ibid, pp. 3-13.

Ibid, p. 14.

Haggett P. (1975), *Geography: A Modern Synthesis,* Harper and Row Publishers, New York, p. 11.

Kirk W. (1975), Problems of Geography, in E. Jones (ed.) *Readings in Social Geography*, Oxford University Press, London, p. 97.

Nag P. (1983), Perception of the Environment, Issues and Challenges, *The National Geographical Journal of India,* Vol. 29, Parts 3 & 4, Varanasi, p. 219.

Ranga K. (1990), Perception of PHC Staff and Delivery of Health Services in Rural Haryana, *The National Geographical Journal of India,* Vol. 36, Part 4, Varanasi, pp. 257-264.

Yeates M. (1974), *An Introduction to Quantitative Analysis in Human Geography*, McGraw Hill Book Company, New York, p. 240.

10

CONCLUSION

Geography is a youthfully growing discipline. It is branching off from its major stems of physical and human geography and the latest off shoot on the human side is behavioural. Even though the interdisciplinary approach is happily acknowledged, the branches of geography draw so heavily, upon other disciplines that it is called 'Jack of all trades'. This happens because geographers go far away from their own field which is the earth's space and appear to have become trespassers. Now, the implications of spatial pattern and location have been clearly recognized as the geographer's particular province. May geography evolve into an elaborate and refined discipline, it is concerned basically with location or space and implications of space. Environment and environmental phenomena that exist in space are bounded by time and space. The first is history, the other geography. Study of phenomena in terms of space or spatial dimensions is geography.

The spatial essence of geography has been aptly grasped by the locationalists and they are developing spatial themes and models for geographical analysis of phenomena. Thus, the locational theory was revived in geography even though it first came from the field of spatial economics, rather than from within itself. Such theorizing is vital for any science or discipline to grow. The locational theories built in geography its solid scientific foundation and also gave it its own identity as a useful and meaningful discipline.

Spatial efficiency is a part and parcel of the locational theory, giving a new insight of the spatial behaviour of man. To be efficient in space is the basic tendency of human beings and therefore the principle of spatial efficiency controls his spatial activities. As a controlling principle of human activities in space, spatial efficiency has a great scope to develop as a fundamental theory and an analytical tool in geography. The classical models like those of Von Thunen, Christaller, Losch, Weber etc. did have an element of spatial efficiency, though not so called. As an eclectic, Isard has synthesized the best of Weber, Losch and Thunen and has demonstrated that location theory and trade theory are but mutually related aspects of a general theory of space economy. What the economists call space economy, is close to the concept of spatial efficiency. It has been very recently recognized by the name of spatial efficiency and as a significant spatial theme.

However, space economy puts the stress on distance minimisation as the prominent localising factor. Geographers, promulgating the idea of spatial efficiency follow the same foot steps, working out the problems on the basis of distance minimization. In this area of locational analysis, the American geographers are pursuing work on spatial patterns of location and spread, spatial structures and systems, density and geometry of flow linkages and using normative models leading to efficiency solutions. In most of the solutions, the basic geographical factor is distance. The concept of spatial efficiency is wider than that. Theoretically speaking, it has two factors to consider at one and the same time-distance minimization is, of course, one of them and the other is the maximizing factor—the maximization of spatial activity. The law of efficiency demands that there should be a comparison or a ratio between maximum and minimum factors *i.e.* maximization of work with minimization of effort. Industrial efficiency has become a watchword in industries to strive for maximum production with minimum waste. Time and motion studies which are popular for efficiency, follow the principle of maximum work with minimum loss of time. In spatial sense, maximum spatial activity with minimum distance which saves effort in space friction or its equivalent like transportation cost, would lead to spatial efficiency. Thus, spatial efficiency is necessarily a maximum-minimum solution and to consider only one of the two factors unless the other is reasonably held constant, is to find an imperfect solution. Sometimes, it is misleading also e.g. the completely connected network of routes is not spatially efficient, though it is the best for the user as it is the minimum distance solution. But it increases the cost of the builder and decreases the utility of the routes which is the maximizing factor. So, the spatial efficiency goes down. Really optimum solution lies in the compromise between both maximum and minimum factors.

The principle of spatial efficiency can be usefully expressed in an operational model which is a simple construction, seeking a ratio between its two maximizing and minimizing variables, and which can be employed in practical application to various geographical situations. It is useful for measurement of the level of spatial efficiency attained by man and also for comparing the levels among themselves. The model may be moulded to suit the purpose and problem at hand but keeping the relation between its two parameters of measurement intact. Von Thunen's model actually applies spatial efficiency in agricultural land use but in a different form. It has based the land rents on distance from the central market and maximization of land use is sought according to the land rents.

Man, with an inherent tendency, strives for spatial efficiency, but he may not reach it perfectly well and there are different levels of its attainment. There are, no doubt, various constraints on him but the main reason is that he does not perceive the factors of spatial efficiency correctly, like the urban people in the Indian cities. That is why his perception of spatial efficiency is important to understand his level of its attainment better.

Spatial efficiency is not only an abstract idea, but also a good measure of spatial organization of human activities also with a scope for prediction and simulation for optimum behaviour. The operational model of spatial efficiency can well be extended to its simulation to find guidelines of efficient solution. The model has thus, the usefulness in planning for development, as it can simulate the optimum. In this sense, the model has the normative value. In the present work, the measurement and simulation of spatial efficiency has been done in a few geographical problems.

More refinement of the basic model of spatial efficiency especially of its simulation and more studies of the areas of its application is a further research. More such models like that of spatial efficiency are also earnestly required to enrich the field of geography.

BIBLIOGRAPHY

Abler R. and J.S. Adams, (1971), *Spatial Organization: A Geographer's View of the World*, Prentice-Hall Inc., Englewood Cliffs, New Jersey.

Ackerman E.A., (1963), Where is a Research Frontier, *Annals of the Association of American Geographers,* 53, 429-40.

Adsul R.S., (2002), Urban Hierarchy of Cities in Maharashtra: A Post Independence Study, *The Deccan Geographer,* Vol. 40, No.2, Pune.

Alcaly R.E., (1974), Aggregation and Gravity Models: Some Empirical Evidence, in M.E.E. Hurst, (ed.) *Transportation Geography*, McGraw-Hill Book Company, New York.

Alexander J.W., (1958), Location of Manufacturing: Methods of Measurement, *Annals of the Association of American Geographers,* 48, 20-26.

——, (1977), *Economic Geography,* Prentice-Hall of India, Pvt. Ltd., New Delhi.

Ambrose P., (1977), *Analytical Human Geography,* Longman, London.

Bairagi S.I and Sapatarshi P.G. (1995), The Impact of Export of Crape in Niphad Tahsil—A Geographical Analysis, MAEER'S, MIT Pune Journal, Vol III, No.12, pp. 40-53.

Barrows H.H., (1923), Geography as Human Ecology, *Annals of the Association of American Geographers,* 13, 1-14.

Bartholomew H., (1955), *Land use in American Cities*, Cambridge, Harvard University Press.

Beckmann M.J., (1952), A Continuous Model of Transportation, *Econometrica*, 20, 643-60.

Berry B. J.L. and W.L. Garrison, (1958-B), A Note on Central Place Theory and the Range of a Good, *Economic Geography,* 34, 304-11

——, (1958-B), Functional Bases of the Central Place Hierarchy, *Economic Geography,* 34, 145-54.

Boyce R.B. and W.A.V. Clark, (1964), The Concept of Shape in Geography, *Geographical Review,* 54.

Bunge. W., (1962), *Theoretical Geography,* Lund Studies in Geography Series C, General and Mathematical Geography, I.

Burgess E.W., (ed.), (1926), *The Urban Community,* The University of Chicago Press, Chicago.

Canter D., (ed.), (1975), *Environmental Interaction,* Surrey University Press, London.

Chapman K. and D. Walker, (1987), *Industrial Location*, Basil Blackwell, New York.

Chatterjee M.K., (1964), An Input-Output Study of the Calcutta Industrial Region, *Papers of the Regional Science Association,* 13.

Chisholm M., (1962), *Rural Settlement and Land use*, Hutchinson University Press, London.

Chorley R.J. and P. Haggett, (1962), *Integrated Models in Geography,* Methuen, London.

Christaller W., (1966), *The Central Places of Southern Germany,* Translated by C.W. Baskin, Prentice-Hall, Englewood Cliffs, New Jersey.

Clark C., (1967), *Population Growth and Land use,* St. Martin's Press, New York.

Cole. J., (1987), *Development and Underdevelopment,* Methuen, London.

Dacey M.F., (1962), *Analysis of Central Place and Point Patterns by a Nearest Neighbour Method*, Lund Studies in Geography, Series B, Human Gegraphy, 24, 55-75.

Dahiya B., (ed.), (1991), *Theoretical Foundations of Development Planning, Vol.3: Sectoral and Regional Planning,* Concept Publishing Company, New Delhi.

Daniels P.W., (1985), *Service Industries: A Geographical Appraisal*, Methuen, New York.

Davis D.H., (1948), *Earth and Man,* The McMillan Co. Ltd., New York.

De Blij H.J. and P.O. Muller, (1985), *Geography: Regions and Concepts,* 4th ed., JohnWiley & Sons, New York.

Deshpande C.D., (1986), Urban Models and Indian Realities, *Transactions, Institute of Indian Geographers,* Vol. 8, No.2.

Dikshit K.R., (1973), Agricultural Regions of Maharashtra, *Geographical Review of India*, Vol. 35, No. 4, pp. 384-396.

Dube S.C., (1974), *Contemporary India and its Modernization*, Vikas Publishing House, Pvt. Ltd., New Delhi.

Dunn E.S., (1954), *The Location of Agricultural Production*, University of Florida Press, Cainesville.

Elhance D.N., (1975), *Fundamentals of Statistics*, Kitab Mahal, Allahabad.

Ericson R.E., (1991), The Classical Soviet-type Economy: Nature of the System and Applications for Reform, *Journal of Economic Perspectives*, 20-23.

Fielding G.J., (1974), *Geography as Social Science*, Harper and Row Publishers, New York.

Fisher J.S., (1992), *Geography and Development: A World Regional Approach,* McMillan, New York.

Friedmann J., (1966), *Regional Development Policy: A Case Study of Venezuela,* The MIT Press, Cambridge.

Friedrich C.J., (1929), *Alfred Weber's Theory of the Location of Industries,* Chicago.

Gadgil D.R., (1965), *Sholapur City: Socio-Economic Studies*, Gokale Institute of Politics and Economics, Pune.

Garrison W.L., (1959-60), Spatial Structure of the Economy, *Annals of the Association of American Geographers*, 49 & 50.

——, (1962), Towards Simulation Models of Urban Growth and Development, *Lund Studies in Geography, Series B, Human Geography,* 24, 92-108.

Getis A., (1963), The Determination of the Location of Retail Services with the use of a Map Transformation, *Economic Geography,* 39, 1-22.

Godlund S., (1957), Bus Service in Sweden, *Lund Studies in Geography, Series B, Human Geography,* 17.

Gould P.R., (1960), Man Against his Environment: A Game-Theoretic Framewwork, *Annals of the Association of American Geographers,* 53, 290-97.

Greenhut M.L., (1956), *Plant Location in Theory and Practice: The Economics of Space*, Chapel Hill, University of N. Carolina Press.

Gregor H.L., (1970), *Geography of Agriculture: Themes in Research*, Prentice-Hall, Inc, Englewood Cliffs, New Jersey.

Griffin E. and L. Ford, (1980), A Model of Latin American City Structure, *Geographical Review,* 70, 397-422.

Grotewald A., (1959), Von Thunen in Retrospect, *Economic Geography,* 35, 346-55.

Haggett P. and R.J. Chorley, (1969), *Network Analysis in Human Geography,* Edward Arnold, London.

Hall P., (ed.), (1966), *Von Thunen's Isolated State,* translated by C.M. Wartenberg, Oxford, England.

Harris C.D. and E.L. Ullman, (1945), The Nature of Cities, *Annals of the American Academy of Political and Social Science*, 242, 7-17.

Hartshorn T.A., (1990), *Interpreting the City: An Urban Geography,* John Wiley & Sons, New York.

Hartshorne R., (1959), *Perspective on the Nature of Geography,* Rand Mc Nally, Chicago.

Harvey D.W., (1966), Theoretical Concepts and the Analysis of Agricultural Land-use Patterns in Geography, *Annals of the Association of American Geographers,* 56.

Henderson J.V., (1988), *Urban Development: Theory, Fact and Illusion,* Oxford University Press, New York.

Hepworth M., (1990), *Geography of the Information Economy,* Gulliford Press, New York.

Hirschman A.O., (1958), *The Strategy of Economic Development,* Yale University Press, New Haven.

Hoover E.M., (1963), *The Location of Economic Activity,* McGraw Hill Book Company, New York.

Howard J.A., (1969), *Theory of Buyer Behaviour,* John Wiley and Sons, New York.

Hoyle B.S. and R.D. Knowles, (1993), *Modern Transport Geography,* Belhaven Press, London.

Huff D.L., (1960), A Topographic Model of Consumer Space Preference, *Papers and Proceedings of Regional Science Association* 6.

Idrisi, (1992), *A Grid Based Geographic Analysis System,* Graduate School of Geography, Clark University, Worcester.

International Institute for Environment, Development and World Resource, (1987), World Resources, New York.

Isard W., (1956), *Location and Space Economy,* The MIT Press, Cambridge.

J. Beaujeu-Garnier and Chabot, (1967), *Urban Geography*, Longman, Green and Co. Ltd., London.

Janelle D.C., (1974), Spatial Reorganisation: A Model and Concepts, in M.E.E. Hurst (ed.) *Transportation Geography,* McGraw-Hill Book Company, New York.

Jasbir Singh and S.S. Dhillon, (1984), *Agricultural Geography,* Tata McGraw-Hill, New Delhi.

Jones C.F. and G.G. Darkenwald, (1965), *Economic Geography,* McMillan, New York.

Kapadnis N.R., (2002), Impact of Environmental Factors on Cropland use in Nashik District, Maharashtra State, *The Deccan Grographer,* Vol. 40. No.1 Pune.

Kates R.W., (1962), Hazard and Choice Perception in Flood Plain Management, *University of Chicago, Department of Geography Research Papers,* 78.

Kayastha S.L., (1986), *Geography and Environment: Issues and Challenges,* Concept Publishing Company, New Delhi.

Kulkarni M.G. (1974), *Problems of Tribal Development: A Case Study,* Parimal Prakashan, Aurangabad.

Losch A., (1954), *The Economics of Location,* translated by W.H. Woglom and W.F. Stolper, Yale University Fress, New Haven.

Lowenthal D., (ed.), (1965), *Man and Nature, Physical Geography as Modified by Human Action,* Cambridge, Mass.

Mahajan Y.N. and V.S. Phadke, (1986), Industries of Khopoli: An Analysis of Raw Material and Market linkage, *Transactions, Institute of Indian Geographers,* Vol. 8, No. 2.

Martin B.V. and C.B. Warden, (1974), Transportation Planning in Developing Countries, in M.E.E. Hurst (ed.), *Transportation Geography,* McGraw-Hill Book Co, New York.

Mc Carty H.H and J.B. Lindberg, (1966), *A Preface to Economic Geography,* Prentice-Hall, Englewood Cliffs, New Jersey.

Mc Connell C.R. and S.L. Brue, (1993), *Macro-Economics: Principles, Problems and Policies,* McGraw-Hill Book Co, New York.

McNee R.B.,(1959), The Changing Relationships of Economics and Economic Geography, *Economic Geography,* 35, (3).

Meadows D., (1974), *Dynamics of Growth in a Finite World*, Wright Allen Press, Cambridge M.A.

Miller G.R. Jr., (1975), *Living in the Environment Concepts, Problems and Alternatives,* Belmont, Wordsworth.

Minshull R., (1972), *The Changing Nature of Geography,* Hutchinson University Library, London.

Mishra R.P., (ed.) (1992), *Regional Planning: Concepts, Techniques, Policies and Case Studies,* Concept Publishing Company, New Delhi.

Mishra R.P., (1971) Diffusion of Agricultural Innovations—A Theoretical and Empirical Study, *Reasranga,* Mysore

Morrill R.L., (1962), *Simulation of Central Place Patterns Over Time,* Land Studies in Geography, Series B, Human Geography, 24, 109-20.

Mountjoy A.B., (1966), *Industrialization and Underdeveloped Countries,* Hutchinson and Co. (Publishers) Ltd. London.

Mulimani A.A., (2002), Hirerarchical Organization of Market Centres in Raichur District, *The Deccan Geographer,* Vol. 40, No.2, Pune.

Murphy R., (1978), *Patterns on the Earth,* Rand McNally, Chicago.

Neuman J. Von and O. Morgenstern, (1944), *Theory of Games & Economic Behaviour,* Princeton.

Norcliffe G.B., (1975), A Theory of Manufacturing Places, in *Locational Dynamics of Manufacturing Industry,* L. Collins and D.F. Walker (ed.), John Wiley & Sons, New York, 19-59.

Ogden P., (1984), *Migration and Geographical Change,* Cambridge University Press, Cambridge, England.

Pahl R.E., (1967), Sociological Models in Geography, in R.J. Chorley and P. Haggett (ed.), *Socio-economic Models in Geography,* Methuen and Co, Ltd. London.

Paterson J.H., (1976), *Land, Work and Resources: An Introduction to Eco. Geog.* Edward Arnold, London.

Peet R., (1987), *International Capitalism and Industrial Restructuring,* Allen and Unwin, Boston.

Penrose E., (1959), *The Theory of the Growth of the Firm,* Basil Blackwell, Oxford.

Perloff H.S. and L. Wingo Jr., (1968), *Issues in Urban Economics,* John Hopkins University Press, Baltimore.

Pocock D and R. Hadson, (1978), *Images of the Urban Environment,* McMillan, London.

Porter P.W., (1965), Environmental Potential and Economic Opportunities: A Background for Cultural Adaptation, *American Anthropologist,* 67, 409-20.

Prabha K., (1979), *Towns: A Structural Analysis: A Case Study of Punjab,* Inter-India Publications, New Delhi.

Prakasa Rao V.L.S and V.K Tewari, (1982), Urban Population Densities: Models and Correlates, *Geographical Review of India,* 44 (2).

Pred A.R., (1967), Behaviour and Location: Foundations for a Geographic and Dynamic Location Theory, (Part-I), *Lund Studies in Geography,* Series B, 27.

——, (1977), *City System in Advanced Economies,* Hutchinson, London.

Raj Bala, (1987), Changes in Territorial Jurisdiction of Towns: A Case Study of Rajasthan, *Annals of the Association of the National Geographers, India,* Vol.7, no.2.

Reid M.C., (1962), *Housing and Income,* Chicago University Press, Chicago.

Reilly W.J., (1931), *The Law of Retail Gravitation,* Knickerbocker Press, New York.

Reissman L., (1964), *The Urban Process: Cities in Industrial Societies,* Free Press, New York.

Riddell J.B., (1970), *The Spatial Dynamics of Modernization in Sierra Leone: Structure, Diffusion and Response,* Northwestern University Press, Evanston.

Rostow W.W., (1960), *The Stages of Economic Growth,* Cambridge University Press, Cambridge.

Rubenstein J.M., (1992), *The Cultural Landscape: An Introduction to Human Geography,* McMillan, New York.

Rushton G., (1969), Central Place *Theory, Annals of the Association of American Geographers,* 62.

——, (1984), Use of Location-allocation Models for improving the Geographical Accessibility of Rural Services in Developing Countries, *International Regional Science Review,* 9, 217-240.

Schaefer F., (1953), Exceptionalism in Geography: A Methodological Examination, *Annals of the Association of American Geographers*, 43, 226-249.

Sharma B.D., (1984), *Planning for Development,* Prachi Prakashan, New Delhi.

Shastri-Ranade P., (1997), Industry and Environmental Dilemmas: A Study of Fertilizer Industry in India, *The Deccan Geographer,* Vol. 35, No.2.

Sinha R., (1975), *Social Change in Indian Society,* Progress Publishers, Bhopal.

Sita K. and J.E. Brush, (1988), The Structure of Greater Bombay, *Annals of the National Association of Geographers, India,* Vol 8, No.1.

Singh S., (1981), Levels of Input-Output Relationship in the Agriculture of Uttar Pradesh, *The Deccan Geographer,* Vol.19, No.283, Secunderabad.

Smith C.T., (1968), Problems of Regional Development in Peru, *Geography,* 53.

Smith D.M., (1981), *Industrial Location: An Economic Geographical Analysis,* John Wiley & Sons, New York.

——, (1997), *Human Geography: A Welfare Approach,* Edward Arnold, London.

Soundaram S.V. and M. Vijayalaxmi, (1986), Administrative Headquarters in Tamil Nadu, Spatial Dimension, *Transactions, Institute of Indian Geographers,* Vol. 8, No.2.

Stamp. L.D., (1960), *Applied Geography,* Penguin Books Ltd., Harmondsworth, Middlesex.

Surya Kant, (1988), A Geographic Perspective on Location of Headquarters of Administrative Areas in India, *Annals of the National Association of Geographers,* Vol.8, No. 2, New Delhi.

Taaffe. E.J., H.L. Gauthier, and M.E. O'Kelly, (1996), *Geography of Transportation,* Prentice-Hall, New Jersey.

Taylor M.J., (1975), Organizational Growth, Spatial Interaction and Location Decision-making, *Regional Studies,* 9, 313-323.

Teitz M.B., (1968), Toward a Theory of Urban Facility Location, *Papers and Proceedings of Regional Science Association,* 35.

Thomas E.N., (1962), Toward an Expanded Central Place Model, *Geographical Review,* 51.

Thomas F.C., (1975), *The Changing Village,* Thomas, Nelson and Sons Ltd., London.

Thomas R.S. and Patton, (1964), *Focus on Geographic Activity: A Collection of Original Studies,* McGraw-Hill Book Company, New York.

Tidswell W.V., (1976), *Pattern and Process in Human Geography,* University Tutorial Press, London.

Tuan Yl Fu, (1974), *Topophilia: A Study of Environmental Perception*, Prentice-Hall, Englewood Cliffs, New Jersey.

Turner R., (1962), *India's Urban Future,* Oxford University Press, London, Bombay.

Ullman E.L., (1974), Geography as Spatial Interaction, in M.E.E. Hurst (ed.) *Transportation Geography,* McGraw-Hill Book Company, New York.

Unwin D., (1981), *Introductory Spatial Analysis,* Methuen, London.

Vidyanath V., (1986), Agricultural Regionalization of Andhra Pradesh, *Annals of the National Association of Geographers,* India, Vol. VI, No.2, 49-56.

Warntz W., (1957), Contribution Toward a Microeconomic Geography: A Review, *Geographical Review,* New York.

Werner C., (1968), The Role of Topology and Geometry in Optimal Network Design, *Papers and Proceedings, Regional Science Association,* 21.

Wibberley C.P., (1959), *Agriculture and Urban Growth: A Study of the Competition for Rural Land,* Michael Joseph, London.

Wilson A.G., (1970), Inter-regional Commodity Flows: Entropy Maximizing Approaches, *Geographical Analysis,* 2.

Wishwakarma R.K., (1981), *Urban and Regional Planning Policy in India,* Uppal Publishing Company, New Delhi.

Wrigley E.A., (1970), Changes in Philosophy of Geography, in R.J.Chorley and P. Haggett (ed.) *Concepts in Frontiers in Geographical Teaching,* Methuen and Co. Ltd., London.

Yeates M., (1963), Hinterland Delimitation: A Distance Minimizing Approach, *The Professional Geographer,* 15 (6).

Yeates M., (1965), Some Factors Affecting the Spatial Distribution of Chicago Land Values, 1910-60, *Economic Geography,* 41.

Yeman K.A., (1968), *Introducing Statistics: Statistics for the Social Scientists,* 1, Penguin, Baltimore.

Zobler L., (1958), Decision Making in Regional Construction, *Annals of the Association of American Geographers,* 48.

Zucker P., (1970), *Town and Square,* The MIT Press, Cambridge, Mass.

INDEX